Advances of DNA Computing in Cryptography

Edited by
Suyel Namasudra and Ganesh Chandra Deka

CRC Press
Taylor & Francis Group
Boca Raton London New York

CRC Press is an imprint of the
Taylor & Francis Group, an **informa** business
A CHAPMAN & HALL BOOK

CRC Press
Taylor & Francis Group
6000 Broken Sound Parkway NW, Suite 300
Boca Raton, FL 33487-2742

Printed on acid-free paper

International Standard Book Number-13: 978-0-8153-8532-5 (Hardback)

Library of Congress Cataloging-in-Publication Data

Names: Namasudra, Suyel, editor. | Deka, Ganesh Chandra, 1969- editor.
Title: Advances of DNA computing in cryptography / edited by Suyel Namasudra, Ganesh Chandra Deka.
Description: Boca Raton : CRC Press, 2019. | Includes bibliographical references and index.
Identifiers: LCCN 2018018628| ISBN 9780815385325 (hardback) | ISBN 9781351011419 (ebook)
Subjects: LCSH: Computer security. | Data encryption (Computer science) | Cryptography--Data processing. | Molecular computers.
Classification: LCC QA76.9.A25 A378 2019 | DDC 005.8/24--dc23
LC record available at https://lccn.loc.gov/2018018628

Visit the Taylor & Francis Web site at
http://www.taylorandfrancis.com

and the CRC Press Web site at
http://www.crcpress.com

Contents

Preface

Secure data communication is a big challenge for information technology (IT) companies due to the increasing number of attackers and malicious users. The exponential increases in users or customers on the Internet makes the situation more critical. Cryptography is the technique of transmitting data or messages (files) between a sender and receiver over an untrusted medium in such a manner that an attacker or malicious user is unable to read the original data. It can also be used for the authentication of customers or users.

DNA can be used to store large chunks of data since large amounts of data can be stored inside the condensed volume of DNA. One gram of DNA can hold approximately 10^{21} bases (A, T, G, and C) of DNA, representing approximately 700 TB of data storage capacity. Therefore, a few grams of DNA can store all the data in the world. The processing speed of write and read operations in DNA computing is comparatively very high, which has already attracted many researchers to use DNA in computing, also known as *DNA computing*.

Many researchers have proposed many schemes for data security using DNA computing. This book starts with a discussion on the technology used in DNA computing in Chapter 1 and gradually introduces the application of various aspects of DNA computing.

Chapter 2 is about the applications of DNA computing in cryptography for data protection, while Chapter 3 deliberates upon the taxonomy of DNA based encryption models. Chapter 4 is a case study on a data encryption scheme using DNA computing, which represents a data encryption scheme for improving data security.

Nowadays, cloud computing is being used in many organizations. Chapter 5 of this book deliberates upon the prospects of DNA computing in the cloud computing environment. Chapter 6 is an analytical study of the taxonomy of security attacks, which is further deliberated upon in Chapter 7. Chapter 8 is a comparative study on various DNA computing algorithms, while Chapter 9 concludes the book by discussing future research trends in DNA computing.

Suyel Namasudra
Ganesh Chandra Deka

Editors

Suyel Namasudra earned his BTech degree in Computer Science and Engineering from the Institute of Engineering and Technology, Uttar Pradesh, India, in 2012. He earned his MTech degree in Computer Science and Engineering from Tripura University (A Central University), Tripura, India, in 2014. Currently, he works as an assistant professor at Galgotias University, Uttar Pradesh, India. His research interests include DNA computing, information security, cloud computing, Big Data, and distributed computing.

Ganesh Chandra Deka is the deputy director (training) in the Directorate General of Training, Ministry of Skill Development and Entrepreneurship (formerly DGE&T, Ministry of Labor & Employment), Government of India. He is a member of the IEEE and IETE, an associate member of the Institution of Engineers (India), and the editor-in-chief of the *International Journal of Computing, Communications, and Networking*. His research interests include information and communications technology (ICT) in rural development, e-governance, cloud computing, data mining, NoSQL databases, and vocational education and training. He has edited three books on cloud computing and published more than 40 research papers. He earned his PhD in computer science from Ballsbridge University, Dominica.

Authors

Bibhudendra Acharya is an assistant professor in the Department of Electronics and Telecommunication Engineering, National Institute of Technology, Raipur, Chhattisgarh, India. His research areas of interest are cryptography and network security, microcontrollers and embedded systems, signal processing, and soft computing. He has authored more than 50 research publications in national/international journals and conferences. He is a member of the editorial board and reviewer for various journals of repute, including those of IEEE and international conferences.

Balusamy Balamurugan is a professor in the School of Computing Science and Engineering, Galgotias University, Uttar Pradesh, India. His main research areas are Big Data, network security, and cloud computing.

Rohan Bali is pursuing a BTech degree in computer science and engineering from Galgotias University, Uttar Pradesh, India. His research interests include DNA computing, information security, and cloud computing.

Sandeep Choudhary earned his BTech degree in Computer Science and Engineering from Guru Gobind Singh Indraprastha University, Delhi, India, in 2008. He earned his MS degree in Cyber Law and Cyber Security from National Law University, Jodhpur, India, in 2012. Currently, he is pursuing a PhD in Computer Science and Engineering from the Birla Institute of Technology, Mesra, Ranchi, India. His research interests include DNA computing, cloud computing, cyber security, web technologies, and Big Data.

Satya Ranjan Dash is an associate professor in the School of Computer Applications, KIIT University, Bhubaneswar, India. He earned his MCA degree from Jorhat Engineering College, Dibrugarh University, Assam, and his MTech degree in Computer Science from Utkal University, Odisha. He earned his PhD in Computer Science from Utkal University, Bhubaneswar, Odisha. His research interests include machine learning, bioinformatics, and cloud computing.

Debashree Devi is a research scholar in the Computer Science and Engineering Department of the National Institute of Technology, Silchar, Assam, India. She earned her BTech degree from Gandhi Institute of Engineering and Technology, Orissa, India, and her MTech degree from Assam Don Bosco University, Assam, India. She has published five research papers in international journals and conferences, including three SCI/SCIE indexing journals. Her areas of interest include DNA computing, Big Data, cloud computing, information security, data mining, artificial intelligence, and pattern recognition.

Suresh Kallam earned his bachelor's and master's degrees from Jawaharlal Nehru Technological University, Hyderabad, and his PhD from VIT University, Vellore, Tamilnadu. He is currently working as a professor, division head, and program chair for MTech, the School of Computing Science and Engineering, Galgotias University, India. Previously, he worked as foreign faculty in East China University of Technology, ECIT Nanchang Campus, Jiangxi, China, and visiting faculty for Jiangxi Normal University,

China. He has published more than 35 national and international conference papers and 25 international journals. He received the Young Scientist Award, Best Faculty Award, and Best Paper Award in 2005 and got first prize in the National Paper Presentation, 2008. His field of interest is the Internet of Things, Big Data, and high performance computing.

Selvaraj Karthikeyan was born in Salem, Tamil Nadu, India, in 1990. He earned his BE (CSE) in 2012 from Anna University, Coimbatore, India, and his ME degree in Computer Science and Engineering in 2013 from Anna University, Coimbatore, India. In 2015, he joined the Department of Computer Science and Engineering at the Knowledge Institute of Technology as an Assistant Professor, and he has been Assistant Professor – G1 at Galgotias University since September 2017. His research interests are cloud computing and security.

Awanish Kumar is an assistant professor at the Department of Biotechnology, National Institute of Technology (NIT), Raipur (CG), India. He completed his PhD at the Central Drug Research Institute (CSIR-CDRI), Lucknow, India, and Jawaharlal Nehru University, New Delhi, India. After the completion of his PhD and before joining NIT Raipur, he served as a post-doctoral fellow at McGill University, Canada, and as UGC Young Research Investigator at Jawaharlal Nehru University, New Delhi, India. Dr. Kumar's main research area is disease biology and health care. His research is focused on molecular/cell biology, computational biology, proteomics, and therapeutic aspects of disease. Dr. Kumar has served in various national and international organizations with different academic/research capacities. He has published more than 75 research papers in Science Citation Indexed (SCI) journals. He is the author of several books/chapters. Currently, he serves on various international/national committees, scientific societies, and advisory panels. He is a member of many international/national professional research societies and reviewer/editorial board member of reputed and refereed international journals. He is regularly invited by top-notch journals to review research papers, and he has reviewed more than 50 research papers.

Rizwan Patan is from Andhra Pradesh, India. He is currently an assistant professor in the School of Computing Science and Engineering, Galgotias University, Uttar Pradesh, India. He completed his PhD in 2017 at the School of Computer Science and Engineering, VIT University, Tamil Nadu, India, and a BTech and MTech in 2012 and 2014, respectively, at Jawaharlal Nehru Technological University Anantapur, India. His publications include four articles in SCI journals and 20 free Scopus indexed journals, conferences, book chapters, and books. He is a guest editor for the *International Journal of Grid and Utility Computing*, *Recent Patents on Computer Science*, and *Recent Patents on Engineering*.

Kintaali Abhimanyu Kumar Patro is a full-time PhD research scholar at the National Institute of Technology Raipur, Chhattisgarh, India. His PhD research area is cryptography and network security. He has six research publications in international journals and conferences. He is a reviewer of the *IEEE Access* journal.

Rahul Roy is co-founder and technical head of ei2 Classes and Technologies, Durgapur, West Bengal, under the approval of MSME and Skill India. His previous assignments have included GATE Faculty, The Gate Academy, Kolkata and Durgapur, India (2012–2015); Technical Assistant, Bengal College of Engineering and Technology, Durgapur (2011–2012); and Service Engineer at Spinco Biotech Pvt Ltd, Kolkata, India (2008–2011). His research interests include medical electronics, embedded systems, digital image processing, speech

recognition, networking tools and IoT. He has published many research papers in various national and international conferences and international journals. He is a member of the Indian Science Congress and VIPNET Club, Government of India.

Alo Sen is founder and director of ei2 Classes and Technologies, Durgapur, West Bengal, India, under the approval of MSME and Skill India. She earned an MTech in Computer Science and Technology and an MCA from KIIT University in 2015 and 2013, respectively. Her research interests include cryptography, steganography, grid computing, cloud computing, and IoT. She has published many research papers in various international conferences and international journals of repute, including IEEE. She is a member of the Indian Science Congress and VIPNET Club, Government of India.

Contributors

Bibhudendra Acharya
Department of Electronics and Telecommunication Engineering
National Institute of Technology
Raipur, Chhattisgarh, India

Balusamy Balamurugan
School of Computing Science and Engineering
Galgotias University
Greater Noida, Uttar Pradesh, India

Rohan Bali
School of Computing Science and Engineering
Galgotias University
Greater Noida, Uttar Pradesh, India

Sandeep Choudhary
Department of Computer Science and Engineering
Birla Institute of Technology
Mesra, Jharkhand, India

Satya Ranjan Dash
School of Computer Applications
Kalinga Institute of Industrial Technology
Bhubaneswar, Odisha, India

Ganesh Chandra Deka
Directorate General of Training
Ministry of Skill Development and Entrepreneurship
Rafi Marg, New Delhi, India

Debashree Devi
Department of Computer Science and Engineering
National Institute of Technology Silchar
Assam, India

Suresh Kallam
School of Computing Science and Engineering
Galgotias University
Greater Noida, Uttar Pradesh, India

Selvaraj Karthikeyan
School of Computing Science and Engineering
Galgotias University
Greater Noida, Uttar Pradesh, India

Awanish Kumar
Department of Biotechnology
National Institute of Technology
GE Road, Raipur, Chhattisgarh, India

Suyel Namasudra
School of Computing Science and Engineering
Galgotias University
Greater Noida, Uttar Pradesh, India

Rizwan Patan
School of Computing Science and Engineering
Galgotias University
Greater Noida, Uttar Pradesh, India

Kintaali Abhimanyu Kumar Patro
Department of Electronics and Telecommunication Engineering
National Institute of Technology
Raipur, Chhattisgarh, India

Rahul Roy
ei2 Classes and Technologies
Durgapur, West Bengal, India

Alo Sen
ei2 Classes and Technologies
Durgapur, West Bengal, India

1

Introduction of DNA Computing in Cryptography

Suyel Namasudra and Ganesh Chandra Deka

CONTENTS

1.1 Introduction

The security of data has become more and more important as data are shared through the cloud in heterogeneous devices (Deka and Das, 2014; Namasudra, 2017, 2018; Namasudra et al., 2017a; Namasudra et al., 2014). *Cryptography* is the technique of secret writing (Willett, 1982; Lin, 1998). The main goal of cryptography is to transmit data or messages (files) between the sender and receiver over an *untrusted* medium in such a manner that an attacker or malicious user is unable to read the original data content. It can also be used for the authentication of customers and users.

In cryptography, the *plaintext* refers to an ordinary readable data or message. The *encryption* process takes plaintext as input and generates *ciphertext* as output by using a secret key and encryption algorithm. The process of recovering the plaintext from the ciphertext by using the secret key is known as *decryption*. Fundamentally, there are two types of cryptosystems, based on how the encryption and decryption processes are carried out:

1. Symmetric key encryption
2. Asymmetric key encryption

The main dissimilarity between these two cryptographic systems is the use of the decryption and encryption keys. Both these keys are strongly associated in any cryptosystem.

In *symmetric* key encryption, the secret key is shared between the sender and receiver, and the encryption and decryption processes are executed by using the same key (Kumar and Wollinger, 2006; Schneier, 1996). Here, the length of the key is small, so the encryption and decryption processes are fast and use less computer processing power. In the 1970s, all cryptosystems used symmetric key encryption. Nowadays, it is still used in many cryptosystems. Some examples of symmetric key encryption are Digital Encryption Standard (DES), IDEA, BLOWFISH and Triple-DES (3DES). There are two main types of symmetric key encryption:

1. *Stream cipher*, where the bytes or digits of a message are encrypted one at a time.
2. *Block cipher*, where a number of bits are taken as a single unit, and then this unit is encrypted. Here, plaintext is padded to make it multiple block size. In block cipher, 64-bit blocks are commonly used.

However, symmetric encryption requires establishing a secure key-exchange mechanism, and the sender and receiver have to trust each other. Otherwise, they may lose their secret information, and there is always a chance that the key will be given to others. Thus, data may face security issues. Figure 1.1 shows the symmetric key encryption process.

Asymmetric key encryption is also known as *public key* cryptography. Here, different keys or values are used to encrypt and decrypt data or a message, and the process does not require sharing keys between the sender and receiver. The keys are usually large numeric numbers and are mathematically correlated to each other even if they are different. The asymmetric key encryption process was invented in the 20th century because it was necessary to pre-share the secret key between two parties. In this process, every party has a pair of keys:

1. A private key
2. A public key, where one key is utilized at the time of encryption, and the other is utilized at the time of decryption (Katayangi and Murakami, 2001).

Computationally, it is unfeasible that one key could be determined based on another key even though both keys are mathematically correlated, which is the strength of asymmetric

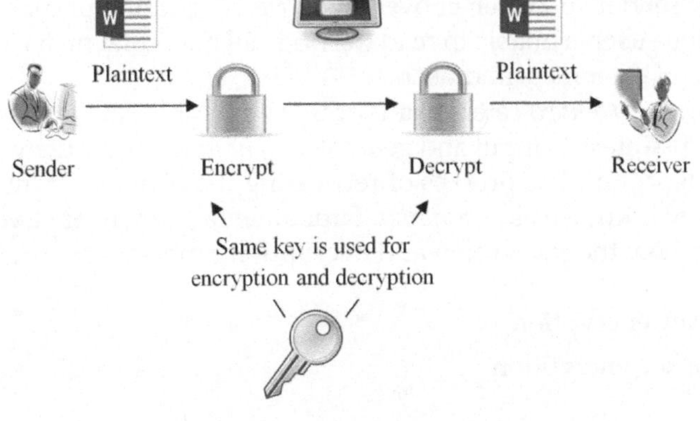

FIGURE 1.1
Symmetric key encryption.

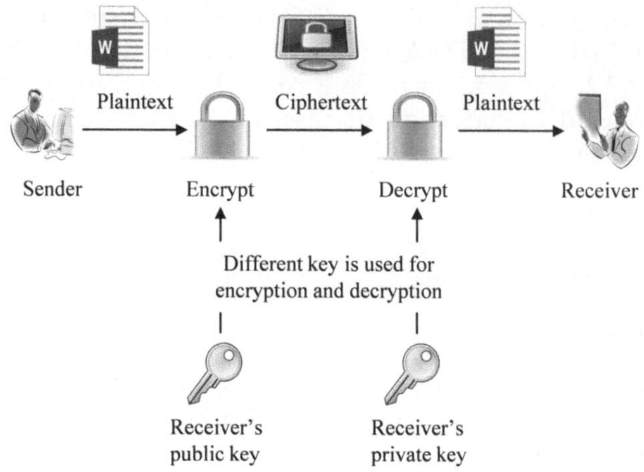

FIGURE 1.2
Asymmetric key encryption.

key encryption. In addition, the strength of the encryption is associated with the key size, and doubling the key length supports an exponential enhancement of its strength. However, users need to trust the third party for key management, which is one of the challenging issues of asymmetric key encryption. Figure 1.2 shows a block diagram of asymmetric key encryption.

Some examples of asymmetric key encryption are Secure Shell (SSH), OpenPGP and digital signature. It is also utilized in many online services, such as browsers that require the establishment of a secure or trusted connection over an unsecured or untrusted network.

To create a better cryptosystem, it is preferable to use symmetric algorithms, since they are fast and secure compared to asymmetric algorithms (Knudsen, 1994; de Cannière et al., 2006).

1.1.1 Key Management in Cryptography

Key management is one of the important aspects of any cryptosystem. There are three main steps in key management:

1. *Key generation*: Key is *generated* for encryption and decryption.
2. *Key revocation*: Key is *revoked* from a user or customer, when s/he is not an authorized user.
3. *Key distribution*: Key is *distributed* in a secure way (Chaeikar et al., 2010). This is the most vital step for key management, since in this step, attackers or malicious users try to get the secret key by using an attack.

So, if a strong algorithm is used for key generation, but the key distribution process lacks security, then the security of the whole cryptosystem might be at risk. An algorithm has been proposed to distribute the key in an unsecured medium (Diffie and Hellman, 1976).

The main contributions of this chapter are to present an overview of cryptography and DNA computing. Here, all the basics of cryptography are presented. Background studies of DNA and the applications of DNA computing in the field of cryptography are also discussed in this chapter.

The rest of the chapter is organized into different parts. Section 1.2 discusses steganography. Background studies of DNA are presented in Section 1.3. DNA computing is discussed in Section 1.4. Applications of DNA computing in cryptography are presented in Section 1.5. Future work directions are discussed in Section 1.6. Finally, Section 1.7 concludes the whole chapter.

1.2 Steganography

Steganography is a technique for hiding data. Both steganography and cryptography have common aim but use different approaches. Table 1.1 summarizes the differences between cryptography and steganography.

Simmons' prisoner's problem is one of the popular illustrations of steganography. According to this example, two criminals want to escape from jail, and they want to discuss their plan without attracting the attention of the police or guards. The only possible way is to communicate with each other by transferring the message or information in a hidden way (Lenti, 2000).

Each method involved in steganography contains an algorithm, which embeds the data or information into a carrier and develops an identifier function that returns the embedded data or information. The function uses a secret key, and this is only shared with or known to the authorized user or customer. The function recovers and identifies the existence of the data or information by using the secret key. Equation 1.1 shows a generic detail of the steganography technique.

$$\text{Cover}_{\text{medium}} + \text{hidden}_{\text{data}} + \text{key}_{\text{stego}} = \text{stego}_{\text{medium}} \tag{1.1}$$

Here, $\text{Cover}_{\text{medium}}$ is a carrier medium in which data ($\text{hidden}_{\text{data}}$) is hidden, $\text{key}_{\text{stego}}$ is the secret key and $\text{stego}_{\text{medium}}$ is the final output of the steganography process.

The most difficult part of steganography is to embed the data or information in such a manner that the attacker or malicious users or the users of the carriers are unable to get it, and most importantly, that the detection of the embedded message or information is also very difficult to recover. Some important factors of steganography are imperceptibility, capacity and robustness. Each factor is responsible for how the data can be embedded into the carrier, how identification of the embedded information or data can be avoided and the strength of the algorithm that protects the data against damage during the embedding process.

TABLE 1.1

Steganography and Cryptography

Sl. No.	Steganography	Cryptography
1	Hides existence of data or information	Changes the content of data
2	Refers to covering a data or information (Johnson and Jajodia, 1998)	Data or information may not be covered
3	Embeds data in a carrier	Data are not embedded
4	A function is used to identify the existence of data	Data are only encrypted by a secret key

1.3 Background Studies of DNA

Biologically, deoxyribonucleic acid is a lengthy molecule that contains most of the genetic information used for the growth, functioning and reproduction of all organisms (Wikipedia, 2018). All the cells in the human body hold the same Wikipedia, and every characteristic of the cells are dependent on it.

Cells can be described as the basic building blocks of all living beings. The human body is made up of trillions of cells, which provide the structure for our bodies. The body takes nutrients from food and converts the nutrients into energy for doing specialized work. Cells also contain genetic material and make copies of it.

Human cells have many parts, which are responsible for different functions. The major parts of human cells are mentioned below:

1. *Cytoplasm*: The cytoplasm is made up of a jelly-like fluid known as *cytosol* along with the other structures that surround the nucleus.
2. *Cytoskeleton*: The cytoskeleton is a long fibrous network that makes up the structure of the cell's framework. It has many functions, such as cell shape, cell division and cell movement.
3. *Endoplasmic Reticulum (ER)*: The endoplasmic reticulum transports process molecules to their destinations, that is, inside the cell or outside the cell.
4. *Golgi apparatus*: *Golgi apparatus* packages the processed molecules, which are transferred to the other cell by the endoplasmic reticulum.
5. *Lysosomes and peroxisomes*: These organelles are the main center for recycling in the cell; they digest bacteria that invade the cell, rid the cell of them and recycle components of the worn-out cell.
6. *Mitochondria*: Mitochondria are among the complex organelles that convert energy into a form that the cell can use. They contain the genetic material that separates the mitochondria from the DNA in the nucleus.
7. *Plasma membrane*: The outer lining layer of any cell is the plasma membrane. It separates any cell from its own environment and supports materials as they leave and enter the cell.
8. *Ribosomes*: Ribosomes are the organelles that process the genetic instructions of the cell for creating proteins. Ribosomes can drift freely within the cytoplasm.

The DNA molecule contains two main biopolymer strands that twist around each other to form a double helix. Two DNA strands are composed of *nucleotides*. Nucleotides are the building block of each DNA molecule and are made up of nucleic acids. The molecules of the nucleic acid are very complex, holding the code that guarantees the exact ordering of 20 amino acids. Interestingly, DNA is comprised of only four units of nucleotide, and each of those nucleotides contains three units:

1. *Sugar molecule*: Sugar molecules are made up of five carbon sugars known as *deoxyribose*.
2. *Phosphate group*: In phosphate groups, one phosphate atom is generally surrounded by four oxygen atoms.
3. *Nitrogen base*: Nitrogen bases are basically the ring compounds made by the nitrogen and carbon atoms, which are structured like a single or double ring. Not all

the nitrogen bases can pair together to form the base pairs. There are four types of nitrogen bases:

a. Adenine (A)
b. Guanine (G)
c. Cytosine (C)
d. Thymine (T)

When covalent bonds are created among the sugars and phosphates of subsequent nucleotides, chains are formed, and nucleotides can join with other nucleotides, which results in the backbone of sugar-phosphate. The outer edges of DNA are formed by alternating the sugar molecules and phosphate groups. DNA molecules are formulated with two single strands, as shown in Figure 1.3. Here, the two strands turn in opposite directions, forming the structure of the double helix. Inside the helix structure, the nitrogenous bases are positioned like "rungs on a ladder" because of a hydrophobic effect and are stabilized by hydrogen bonding. DNA strands are kept together because of the hydrophobic interactions and hydrogen bonds. The hydrogen bonds are created between the base pairs. The base of the first strand only creates a hydrogen bond with a specific base of the second strand. These two bases form a base pair. A is paired with T and C is paired with G according to the rules of base pairing. Large numbers of DNA sequences can be generated by using the combinations of random DNA bases. The National Center for Biotechnology Information (NCBI) database gives a huge number of sequences of DNA bases (NCBI, 2018).

In DNA, *codons* can be defined as a group of three neighboring nucleotides that corresponds to any amino acid or stop signal at the time of protein synthesis. DNA sequences or molecules are written by using four nucleotides, and a protein involves 20 amino acids. Codons support the key that allows DNA and protein to translate into each other. Each codon usually corresponds to a single amino acid. The full set of codons is known as the *genetic code*. Table 1.2 shows the genetic code (DNA). Here, the genetic code involves 64 possible combinations or permutations of the sequence of the three-letter nucleotide, which can be taken from the four nucleotides. Here, 61 out of 64 codons represent amino acids, and the other three represent stop signals. For example, the codon TAA represents a stop codon, and the codon CAG represents the amino acid, namely Glutamine (Gln). The genetic code can be referred to as *redundant* or *degenerate* because one single amino acid can be used at the time of coding for more than one codon.

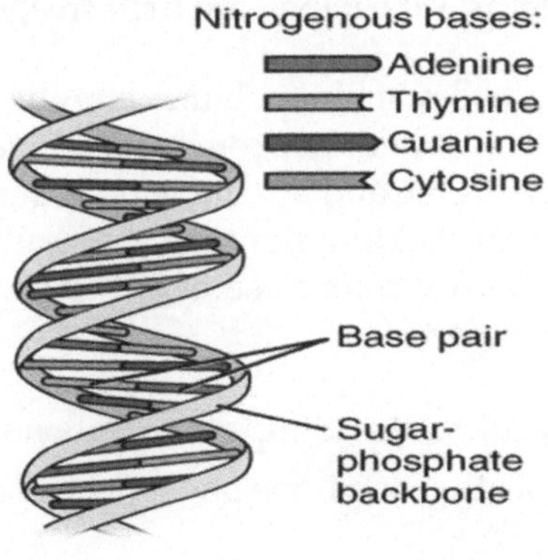

FIGURE 1.3
DNA structure.

TABLE 1.2
The Genetic Code (DNA)

Codon	Amino Acid	Codon	Amino Acid	Codon	Amino Acid	Codon	Amino Acid
TTT	Phe	TCT	Ser	TAT	Tyr	TGT	Cys
TTC	Phe	TCC	Ser	TAC	Tyr	TGC	Cys
TTA	Leu	TCA	Ser	TAA	STOP	TGA	STOP
TTG	Leu	TCG	Ser	TAG	STOP	TGG	Trp
CTT	Leu	CCT	Pro	CAT	His	CGT	Arg
CTC	Leu	CCC	Pro	CAC	His	CGC	Arg
CTA	Leu	CCA	Pro	CAA	Gln	CGA	Arg
CTG	Leu	CCG	Pro	CAG	Gln	CGG	Arg
ATT	Ile	ACT	Thr	AAT	Asn	AGT	Ser
ATC	Ile	ACC	Thr	AAC	Asn	AGC	Ser
ATA	Ile	ACA	Thr	AAA	Lys	AGA	Arg
ATG	Met (Start)	ACG	Thr	AAG	Lys	AGG	Arg
GTT	Val	GCT	Ala	GAT	Asp	GGT	Gly
GTC	Val	GCC	Ala	GAC	Asp	GGC	Gly
GTA	Val	GCA	Ala	GAA	Glu	GGA	Gly
GTG	Val	GCG	Ala	GAG	Glu	GGG	Gly

A *primer* can be defined as ten sequence pairs of DNA sequences. It is a short length of nucleic acid sequence, which provides the starting point of the DNA synthesis. A primer is usually synthesized by the enzyme called *primase* before the DNA replication. The enzymes that synthesize DNA are called *DNA polymerases*. DNA polymerases are only able to connect the existing strand of the nucleotides with the new nucleotides. Therefore, the primer must undergo a synthesis process, and the primer serves as the key and foundation of DNA synthesis. Before the completion of DNA replication, primers are removed, and DNA polymerases are used to fill the gaps in the sequence. In the laboratory, DNA primers can be designed and synthesized with the specific sequences that can bind the DNA sequences into a single strand. These designed DNA primers are generally used for performing the Polymerase Chain Reaction (PCR). PCR is a method or process for creating many copies of the DNA sequences, which involve continual reactions by a polymerase.

1.4 DNA Computing

DNA computing is a computing area wherein DNA, molecular biology, hardware and biochemistry are used to encode genetic details in computers. Research and development of DNA computing involves experiments, theory and application.

In 1994, DNA was first proposed in the field of computation by Adleman (1994) of the University of Southern California. Adleman demonstrated a proof-of-concept by using DNA in the form of computation that solved the seven-point "Hamiltonian Path Problem." Initially, this novel approach was used to solve NP-hard problems. However, very soon it was realized that DNA computing may not be the best solution for this type of problem. Computer scientist Mitsunori Ogihara and biologist Animesh Ray described an implementation of Boolean circuits by using DNA computing in 1997 (Ogihara and Ray, 1999). Since

then, many techniques have been developed by using DNA to solve many problems, such as the SAT problem, the 0-1 planning problem, the integer planning problem, graph theory, the optimal problem, database and cryptography, and many Turing machines are proved by using DNA computing. Currently, DNA computing is one of the trending fields in biology and computer science. A significant background in both computer engineering and DNA molecules are essential for developing an efficient algorithm by using DNA computing.

A *computation* can be defined as the implementation of an algorithm, that is, a list of step-by-step instructions, which takes inputs, processes those inputs and gives a result. In DNA computing, rather than the traditional binary digit (1 and 0), data (or information) are mainly represented by using four genetic alphabets:

1. A
2. T
3. C
4. G

It is achievable since the short molecules of DNA of any random sequence can be synthesized. The input of an algorithm is therefore represented by using the DNA molecules with particular sequences, and on the molecules, the instructions are used by the laboratory operations to get the result of the final set of molecules.

DNA computing is one of the forms of parallel computing, and it takes advantage of numerous different DNA molecules (Lewin, 2002). The complexity and organization of all human beings is mainly based on the coding system of the four components of DNA molecule. One strand of DNA consists of four bases (A, T, C and G). After attaching themselves to the deoxyribose, those nucleotides form a string for generating the long sequences (Boneh and Lipton, 1996). By using processes drawn from chemistry and physics, the strands can be separated (Paun et al., 1998). Different DNA structures are used by researchers for solving different problems (Lipton, 1995). Most of the DNA molecules used in DNA computing are double strand and single strand. Some DNA strands used in DNA computing can be hairpin and plasmid DNA molecules. There are many operations in DNA computing models, namely *adjoining*, *cut*, *insertion*, *paste* and *deletion*.

1.4.1 Why DNA Computing?

DNA computing can be used to deal with scientific problems and calculations that cannot be easily solved by using existing techniques of data sharing and storage. DNA computing is popular for three main reasons:

1. *Speed*: Conventional or traditional computers are able to perform approximately 100 Million Instructions Per Second (MIPS). However, combinations of DNA strands have a computational power equivalent to 10^9 or more. In comparison to the fastest computer, DNA is also considered to be 100 times faster.
2. *Minimal storage requirements*: DNA stores memory space at the density of approximately 1 bit per cubic nanometer. Conventional storage systems, on the other hand, need 10^{12} cubic nanometers for storing 1 bit.
3. *Minimal power requirements*: In DNA computing, power is not required at the time of computation. The chemical bonds that are responsible for the construction of DNA do not require a power source.

An important issue in DNA computing is how researchers can minimize the probability of mistakes at the time of execution. Numerous experiments have been carried out to prove that the reliability of DNA computing can be enhanced extensively by using a suitable encoding approach (Boneh et al., 1996). All researchers want to obtain the maximum encoding set, and this can be achieved by using two approaches. The first is to use random arithmetic, which is a common approach. The second is to use construct arithmetic, and this is better than random arithmetic since this approach takes less time to find the right codes.

1.5 DNA Computing in Cryptography

Information security has a significant responsibility in many fields, such as military affairs, financial companies and confidential business. In traditional systems, users are limited to their own domain. However, in the advanced technologies of today's world, users want to access and store data outside of their own domain. So, data security becomes crucial due to the presence of many attackers and hackers who are always trying to hack users' personal and confidential data for their own sake or for earning revenue. Sometimes, they replace the original data with fake data. Thus, again, users' data face a security issue.

Many cryptographic methods have been developed for the transfer of data. Shamir (1985) first proposed *identity-based encryption* (IBE). In this process, an identity is specified by the sender of any data, and this must be matched by the receiver for data decryption. A few years later, the *fuzzy identity–based encryption* model was proposed, wherein the identity of the user is presented with a set of descriptive attributes. A respective user can only decrypt the encrypted data, if the attributes exactly match the ciphertext's attributes. *Attribute-Based Encryption* (ABE) was proposed by Sahai and Waters (2005) for providing complex data accessing. *Ciphertext Policy–based ABE* (CPABE) and *Key Policy–based ABE* (KPABE) are two main types of ABE. In CPABE, ciphertext is assigned to an access policy, and private keys are generated for authorized users on the basis of users' attributes (Bethencourt et al., 2007). An authorized user can decrypt the ciphertext, if the attributes of the user satisfy the assigned access policy in the ciphertext. In KPABE, ciphertext is assigned with the attributes, and an access structure is associated with the user's private key, which denotes the user's identity (Goyal et al., 2006). Here, an authorized user can decrypt the ciphertext, if his/her access tree satisfies all the respective attributes of the ciphertext. Most of the traditional methods are mainly based on tough complex mathematical equations, where researchers have mainly focused on increasing the complexity by modifying the equations (Lin and Hsueh, 2008; Liu and Tsai, 2007; Sencar et al., 2007). In some traditional approaches, to improve security, the data are usually encrypted and the encrypted data embedded into various multimedia data or files that serve as cover media, like audio, video and images. (Voloshynovskiy et al., 2003; Barni et al., 2003). Thus, malicious users or attackers are unable to find the data for breaking the cryptosystems.

Nowadays, DNA cryptography is one of the rapid technologies that is based on DNA computing. Here, DNA computing is used for achieving a strong security mechanism for data encryption and data storage, so that unauthorized, malicious users and attackers are unable to read the data content. The most important aspect of DNA computing technology in the areas of steganography and cryptography is that it can be used to achieve higher levels of trust for unbreakable algorithms.

Many cryptographic algorithms have been proposed based on DNA, namely symmetric and asymmetric key cryptosystems using DNA, triple-stage DNA cryptography, DNA steganography systems, DNA-based chaotic computing and DNA-based encryption algorithm. In a DNA-based encryption scheme, there are a few criteria that need to be fulfilled to achieve strong security:

1. *Complete character for DNA encoding*: A complete character set of DNA encoding must provide numbers, alphabets (uppercase and lowercase) and special characters. This is used to encode all the characters of the plaintext of a message or data into a DNA sequence. In existing models, the mapping table, which provides the complete characters of DNA, is not secured since the table is manually developed (Agrawal, 2012). So, confidential or sensitive data may face security issues. One good solution is to automatically develop the mapping table so it will be very difficult for attackers to hack the data.
2. *Generation of dynamic encoding table*: To provide strong security, the encoding table must be dynamically generated, and it is changed for every session between the sender and receiver of the data. The encoding table must also provide different DNA bases for each element of the plaintext.
3. *Unique sequence to encode each character of the plaintext*: Between the sender and receiver of the data, at the time of generation of the encoding table in each session, the encoding rule or formula for translating a plaintext into the DNA sequence must be unique for each and every element of the character set. Thus, data security can be increased against attackers and malicious users.
4. *Robustness of the encoding process*: In order to achieve security against attacks, the DNA encoding rule or formula of the plaintext must support robustness, which makes it much more difficult to decipher.
5. *Biological simulation process*: DNA-based encryption and decryption processes must be mainly based on biological processes, which are simulated for adaptation to the digital computing environment. In many existing models, encryption and decryption processes are mainly based on biology laboratory experiments, which are not very suitable in the digital computing environment. Since DNA cryptography is used to divide modern cryptographic algorithms into several parts, a complete, well-defined algorithm must be based on the simulation of biological processes.
6. *Dynamicity of the encryption process*: Dynamicity is needed in the data encryption process for ensuring different ciphertext, even for the same plaintext. This can be achieved by combining both DNA cryptography and modern cryptography this can be achieved.

1.5.1 Polymerase Chain Reaction

PCR can be defined as a process of quantification and amplification of DNA. The main purpose of PCR is to increase the quantity of DNA because it is very difficult to carry out computations on a small quantity of DNA strands. The term *polymerase chain reaction* arises from a biological catalyst, which is known as *polymerase*. A short sequence of DNA can be amplified and analyzed by performing PCR. The process of DNA amplification includes the cloning of segments. PCR is very efficient and can generate huge numbers of copies even from a small amount of selected DNA. In addition, PCR utilizes the molecules that are used in nature to copy DNA. The sequence of the DNA must be known in order to

perform PCR, and a suitable primer is needed to design an amplification process. The PCR process comprises two phases:

1. Two primers are related to the ending and beginning of the DNA strand.
2. The polymerase enzyme is moved along the fragment of the DNA strand, reading the code, and copies are assembled.

Here, the encryption key is compound, which keeps both the public-key and PCR-primers pair, and the decryption key keeps private key and complementary-primer pairs. For a data communication process between Alice (sender) and Bob (receiver), two primers are shared between Alice and Bob through a secure channel before starting the encryption process (Cui et al., 2008). In the data encryption process, an algorithm such as Advanced Encryption Standard (AES) or RSA (Ron Rivest, Adi Shamir and Leonard Adleman) can first be applied as the pre-processing step. Then, the ciphertext of the data is converted into the DNA strand by using a coding rule. Thus, a different ciphertext is generated. *Cipher DNA* denotes the ciphertext of the data that is in the form of a DNA sequence, and *plaintext DNA* refers to the plaintext of the data in the form of DNA sequence.

The cipher DNA is flanked by using the primers (secret), and then, it is mixed with other unknown DNA sequences. Alice then sends this mixed DNA to Bob. Bob retrieves the block cipher DNA by running PCR with the help of the secret primer, and then, reverses all the processes that are used for data encryption. Without the prior knowledge of the two primers, no one is able to retrieve the cipher DNA. Figure 1.4 shows the cryptographic method using PCR.

The aforementioned PCR method has various implications; Cui et al. (2008), for example, have proposed a data encryption model based on DNA coding and PCR amplification. Kazuo et al. (2005) have proposed a public-key system using a one-way function based on DNA. By using a large-scale DNA space based on PCR, Yamamoto et al. (2008) have proposed a novel encryption scheme, which provides two level securities by using both the modern algorithms and molecular techniques. If any level of security has been compromised, another level keeps the system safe. However, secret key exchange between the sender and receiver is one of the major problems of PCR.

1.5.2 DNA-Based Steganography

For data encryption in DNA steganography, one or more DNA strands are taken as input in the form of plaintext. With the input DNA, randomly generated strands of secret key are appended. Then, the resulting plaintext DNA sequence is hidden by being mixed with numerous distractor DNA strands, which also might be generated

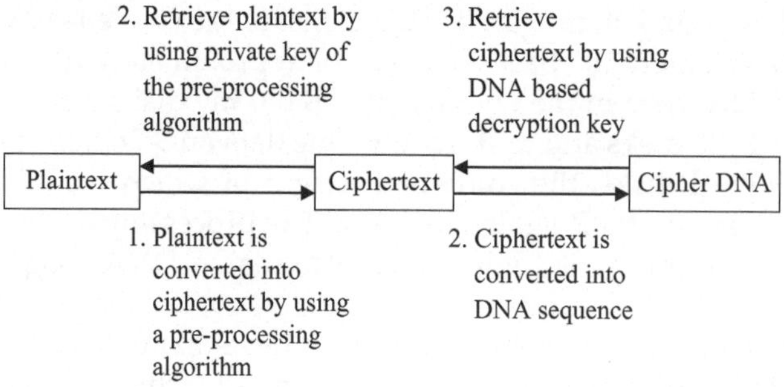

FIGURE 1.4
Cryptographic method using PCR.

randomly. For data decryption, the DNA strands are decrypted by using a possible set of known DNA separation methods with the help of the secret key. Such plaintext message strands can be separated by using the hybridization process with the complements of the secret-key strands. These separation processes can be combined by using amplification steps.

Risca (2001) has proposed a DNA-based steganographic technique based on the use of standard biological protocols. This method encrypts the information or data in the form of DNA sequence and is flanked by using two different secret primers. This encryption technique utilizes a monoalphabetic encryption key for assigning three-base DNA codons. The key is then used to encode the plaintext of the data or messages into the DNA sequence. Then, the resultant DNA sequence is hidden in a large DNA strand. To extract the message or data, an authorized user must know the primers, which combine the target regions to amplify the required molecule.

The DNA-based steganography technique is simple, only converting the plaintext of the data into a plaintext DNA. It does not encrypt any data, and it only conceals the plaintext DNA within other DNA strands. Any user who knows the primers can easily find the location of the plaintext DNA and can easily amplify it for the data decryption process. The DNA-based steganography technique can be efficient since it minimizes the cost of encryption. However, it is considered vulnerable for statistical analysis. Gehani et al. (2000) have proposed an improved method that involves separating the distractor DNA stands and the probability distribution of the plaintext of the data or message.

1.5.3 DNA-Based Chip Technology

DNA chips and microarrays have been used for over a decade and have changed recent technology. DNA chips are enabled to manipulate a large amount of data or messages from genome sequencing (Gwynne and Heebner, 2001). DNA-based chip technology is very significant for the manipulation of any kind of data. It is usually used to find the expressions of numerous genes. These chips are like silicon chips, which are mainly used to store the data or message in the DNA sequence form.

DNA-based chips are mainly made by a large number of embedded spots on the solid surface. Each spot holds *probes*, which can be defined as small nucleotide sequences that can bind into the complementary nucleotides. Nucleotides that can bind these probes are labeled. Whenever a DNA sequence is connected with these probes, in each spot, the data are calculated electronically based on the ratio of the binding of a probe with respect to the DNA sequence (Tsukahara and Nagasawa, 2004). DNA is attractive for data storing since a large number of data items can be stored inside the condensed volume. One gram of DNA holds approximately 10^{21} bases of DNA, or approximately 700 TB. So, a few grams of DNA can store all the data in the world. If 50 GB Blu-ray disks were used, it would take up approximately 14,000 disks and to store the same data on 3 TB hard disks, it would take approximately 233 hard disks, the combined weight of which would be approximately 151 kilograms (Sathappan and Pragaladan, 2016). The processing speed of write and read operations in DNA computing is high, which also makes DNA computing attractive to many researchers.

Nowadays, manufacturers are concentrating on developing small biochips with high information or data-handling capabilities. These small biochips can be very effective in many research fields, including cryptography. There are few steps in these typical cryptographic techniques:

1. In this process, the collection of particular probes is considered as an encryption key, and the collection of respective probes that make up a complimentary sequence are considered as a decryption key. The receiver of the data communication process receives the decryption key in a secure manner.
2. The plaintext of a data is converted into binary format. Then, this binary format is embedded as the cipher DNA into a DNA chip. No one can access the plaintext without the decryption key.
3. The receiver hybridizes the cipher DNA by using the decryption key and can retrieve the plaintext by using any suitable computer software.

Exclusive OR (XOR) one-time-pad is an example of the abovementioned cryptosystem (Gehani et al., 2000). The powerless array of DNA strands is used for constructing one-time-pad by using DNA-based chip technology, where many copies of one DNA sequence are gathered in a pixel. The strands can be addressable optically, and at each and every site, where they are optically addressed, distinct DNA sequences are synthesized. Combinatorial synthesis is a popular technology in this technique, which is conducted at thousands of locations.

To prepare oligonucleotides of length L in $4n$ chemical reactions, 4^L sequences are synthesized. To encrypt the plaintext of a message or data, a prefix unique index of length L_0 is appended. On the one-time-pad DNA sequence, one complement of the plaintext data or message is generated by appending the same length of prefix unique index L_0. A one-time-pad DNA sequence and the respective pair of the plaintext data are concatenated by using ligation and annealing. Then, bit-wise XOR operation is used to encrypt the data. XOR operation is presented in Equation 1.2, where M = plaintext of the data or message, C = cipher data or strands and S = sequence of random bits.

$$C_i = M_i \otimes S_i \tag{1.2}$$

where, $i = 1, 2, \ldots, n$

Here, cipher strands are converted from the segments of the plaintext, and the plaintext strands are dripped. Equation 1.3 shows bit-wise XOR operation for data decryption:

$$C_i \otimes S_i = (M_i \otimes S_i) \otimes S_i = M_i \otimes (S_i \otimes S_i) = M_i \tag{1.3}$$

A similar DNA-based chip technology is used by MingXin et al. (2007) to design a symmetric key encryption model, namely DNA-based Symmetric key Cryptosystem (DNASC), by using DNA computing. Lai et al. (2010) have designed a signature using DNA-based chip technology and have proposed an asymmetric encryption scheme.

The use of DNA-based chip technology for data encryption and decryption can be useful for all kinds of data, including textual and image data (Shyam et al., 2007). However, the issue of the interoperability of storage mediums and DNA chips suppress the potential of DNA-based chip technology.

1.6 Future Research Work Directions

On the basis of the above discussion, there are huge opportunities for the use of DNA computing in several areas:

- A new scheme can be developed by using the *complementary pair rule*. A complementary pair is a unique equivalent pair that is assigned to every nucleotide base pair. A researcher can define his/her own complementary rule by using the DNA bases (A, T, C and G) to increase the complexity of the algorithm.
- In the existing technology, primers called *keys* are added in the DNA sequence. However, a primer can also be selected from the database, and after it has been divided into two parts, primer can be added to both sides of the ciphertext to increase the security of the user's confidential data. Thus, a new scheme can be developed by using a primer.
- In existing schemes, the main emphasis is put on the data security issue. However, the time complexity can be increased because of the many complex operations involved. A new scheme can be developed by using DNA computing, which can increase data security as well as decrease time complexity.
- DNA computing can be used for DNA certification and steganography, which offers more protection than any other encryption technique. A new steganographic technique can be developed by using DNA computing to increase data security.
- In many advanced schemes, users' attributes are used to define a secret key. These secret keys can be converted into DNA bases to improve data security.
- To increase the complexity of the algorithm, a new encryption scheme can be developed by using DNA computing and American Standard Code for Information Interchange (ASCII) value in which, at first, a data is converted into its binary form. Then, this binary form is converted to the DNA bases and these DNA bases are again converted to its corresponding ASCII values. Thus, data security can be increased.
- In the cloudcomputing environment, there are huge amounts of confidential data as well as big data. Due to the presence of many attackers or malicious users, data confidentiality may be at risk (Namasudra and Roy, 2015; 2016; 2017a–2018). A new data encryption scheme can be developed for the cloudcomputing environment to improve data security.
- Application of DNA Computing by 2020 and beyond: Table 1.3 shows the potential applications of DNA computing in different fields by 2020 and beyond.

1.7 Conclusions

DNA computing is widely used to improving data security because of its complex structure. In this chapter, a detailed discussion about cryptography, steganography and the biological background of DNA and DNA computing, as well as of its importance, has been presented. There are many DNA computing technologies in the field of cryptography, namely polymerase chain reaction, DNA-based steganography and DNA-based chip technology. All these technologies have been discussed in detail in this chapter. Moreover, future research work directions are also presented in this chapter, which can be highly beneficial to develop novel DNA computing–based techniques.

TABLE 1.3

Applications of DNA Computing in Different Fields

Strategy/Area	Applications
DNA fountain strategy	As per the report of International Data Corporation firm of Framingham, data will reach 44 trillion gigabytes by 2020. So, these data need to be saved somewhere, and DNA computing can be useful to save these data. Magnetic tapes or hard disks cannot be useful since they corrupt within a decade, but DNA can be steady for millennia without any power required for storage. Nowadays, the DNA fountain strategy is used to encode the data in DNA as Sudoku puzzles. This strategy can store 10 times more data (IDC, 2014).
Online services	Many researchers consider that the DNA storage technique can be useful for online archiving services, where long-term data storage services are required (O'Hare, 2016).
Quantum computing	Extremely short, configurable pulses of light can operate computers 100,000 times faster than today's computers by using quantum computing. So, when these computers become available for commercial use, researchers can easily use DNA computing on those computers for fast processing (Satell, 2016). Recently, companies like Microsoft, IBM and Google have begun investing huge amounts of money in quantum computing (García-Torres et al., 2016).
Cognitive machines	Researchers in South Korea are trying to develop *cognitive machines*, which are based on physical intelligence and the convergence of biology, and they are also trying to design biologically inspired robots based on DNA computing (Chandra, 2017).
Sequence market	By 2020, the market for public cloud computing will have increased to $46 billion (Gartner, 2017). The sequence market is developing day by day using the technology of cloud computing, where the main emphasis is on the applications and new technologies of DNA sequences, such as pyrosequencing, 454 technology, Massively Parallel Signature Sequencing (MPSS) and Sequencing by Synthesis (SBS) (Lin et al., 2011).
Blockchain quantum	The Intel Corporation applied for a patent for "Blockchain System with Nucleobase Sequencing as Proof of Work" in June 2016 to offset some of the inefficiencies in Proof of Work (PoW) cryptocurrency mining. Instead of solving arbitrary cryptographic problems for authentication, miners will be processing DNA sequences for verifying the authenticity of transaction history on a blockchain (Reese, 2017; Raj and Deka, 2018). Sharing sensitive DNA data, blockchain and genomics go well together (Eaves, 2018). EncrypGen stores DNA data on the blockchain. EncrypGen helps people to sell genetic code in return for cryptocurrency (Ramsay, 2018). KuCoin traders will have access to DNA/BTC and DNA/ETH trading pairs upon listing the KuCoin as tradable tokens (Weill, 2018).

Key Terms and Definitions

Attribute-Based Encryption (ABE): In attribute-based encryption, data are encrypted by using a set of attributes to improve data security. Here, only users, who have the valid set of attributes can decrypt the data, and thus, access the original content.

Ciphertext Policy Attribute-Based Encryption (CPABE): In CPABE, the user's private key is generated based on the attributes of the user, and each ciphertext is associated with an access policy. A user or customer can decrypt the ciphertext, when the attributes of the user satisfy the access policy of the ciphertext.

Key Policy Attribute-Based Encryption (KPABE): In KPABE, the user's or customer's private key is assigned to an access structure, and the ciphertext of the data is allied with a set of attributes. The users can decrypt the ciphertext, if their access tree satisfies the ciphertext's attributes.

References

Adleman, L. M. (1994). Molecular computation of solutions to combinatorial problems. *Science*, 266(5187), 1021–1024.

Agrawal, A., Bhopale, A., Sharma, J., Ali, M. S., and Gautam, D. (2012). Implementation of DNA algorithm for secure voice communication. *International Journal of Scientific and Engineering Research*, 3(6), 1140–1144.

Barni, M., Bartolini, F., and Furon, T., (2003). A general framework for robust watermarking security. *Signal Processing*, 83(10), 2069–2084.

Bethencourt, J., Sahai, A., and Waters, B. (2007). Ciphertext-policy attribute based encryption. In *Proceedings of the 2007 IEEE Symposium on Security and Privacy*, IEEE, Berkeley, CA, USA, pp. 321–334.

Boneh, D. and Lipton, R. (1996). Making DNA computers error resistant. In *Proceedings of Second Annual Conference on DNA Based Computers*, pp. 102–110.

Boneh, D., Dunworth C., Lipton, R. J., and Sgall, J. (1996). On the computational power of DNA. *Discrete Applied Mathematics on Computational Molecular Biology*, 71(1–3), 79–94.

de Cannière, C., Biryukov, A., and Preneel, B. (2006). An introduction to block cipher cryptanalysis. *Proceedings of IEEE*, 94(2), 346–356.

Chandra, R. (2017). An affective computational model for machine consciousness. *arXiv preprint arXiv:1701.00349*.

Chaeikar, S. S., Razak, S. A., Honarbakhsh, S., Zeidanloo, H. R., Zamani, M., and Jaryani, F. (2010). Interpretative key management (IKM), a novel framework. In *Proceedings of the Second International Conference on Computer Research and Development*, IEEE, Kuala Lumpur, Malaysia, pp. 265–269.

Cui, G., Qin, L., Wang, Y., and Zhang, X. (2008). An encryption scheme using DNA technology. In *Proceedings of the IEEE 3rd International Conference on BioInspired Computing: Theories and Applications (BICTA08)*, IEEE, Adelaide, SA, Australia, pp. 37–42.

Diffie, W. and Hellman, M. E. (1976). New directions in cryptography. *IEEE Transactions. Information Theory*, IT-22(6), 644–654.

Deka, G. C. and Das, P. K. (2014). An overview on the virtualization technology. *Handbook of Research on Cloud Infrastructures for Big Data Analytics*, IGI Global. DOI:10.4018/978-1-4666-5864-6.ch012.

Eaves, S. (2018). Blockchain the innovation beyond cryptocurrencies. Retrieved February 17, 2018, from http://www.arabianbusiness.com/technology/389540-blockchain-the-innovation-beyond-cryptocurrencies.

García-Torres, M., Gómez-Vela, F., Melián-Batista, B., and Moreno-Vega, J. M. (2016). High-dimensional feature selection via feature grouping: A variable neighborhood search approach. *Information Sciences*, 326, 102–118.

Goyal, V., Pandey, O., Sahai A., and Waters, B. (2006). Attribute-based encryption for fine-grained access control of encrypted data. In *Proceedings of the 13th Conference on Computer and Communications Security*, IEEE, New York, NY, USA, pp. 89–98.

Gehani, A., LaBean, T., and Reif, J. (2000). DNA-based cryptography. In Jonoska, N., Paun, G., and Rozenberg, G. (Eds.), *Aspects of Molecular Computing*. Berlin Heidelberg: Springer, pp. 167–188.

Gwynne, P. and Heebner, G. (2001). Technologies in DNA chips and microarrays. *Science*, 4, 949.

IDC (2014). The digital universe of opportunities: Rich data and the increasing value of the internet of things. Retrieved June 11, 2017, from https://www.emc.com/leadership/digital-universe/2014iview/executive-summary.htm.

Johnson, N. F. and Jajodia, S. (1998). Exploring steganography: Seeing the unseen. *IEEE*, 31(2), 26–34.

Katayangi, K. and Murakami, Y. (2001). A new product-sum public-key cryptosystem using message extension. *IEICE Transactions on Fundamentals* E84-A(10), 2482–2487.

Kazuo, T., Akimitsu, O., and Isao, S. (2005). Public-key system using DNA as a one-way function for key distribution. *BioSystems*, 81(1), 25–29.

Knudsen, L. R. (1994). Block ciphers – analysis, design and applications. PhD thesis, Aarhus University, Denmark, D AIMI PB 485.

Kumar, S. and Wollinger, T. (2006). Embedded security in cars. In Lemke, K., Paar, C., and Wolf M. (Eds.), *Fundamentals of Symmetric Cryptography*. Berlin Heidelberg: Springer, pp. 125–143.

Lai, X., Lu, M., Qin, L., Han, J., and Fang, X. (2010). Asymmetric encryption and signature method with DNA technology. *Science China Information Sciences*, 53(3), 506–514.

Lin, H. S. (1998). Cryptography and public policy. *Journal of Government Information*, 25(2), 135–148.

Lin, C. C. and Hsueh, N. L. (2008). A lossless data hiding scheme based on three-pixel block differences. *Pattern Recognition*, 41(4), 1415–1425.

Lin, B., Wang, J., and Cheng, Y. (2011). Recent patents and advances in the next-generation sequencing technologies. Retrieved June 11, 2017, from https://www.ncbi.nlm.nih.gov/pmc/articles/PMC3122325/.

Lipton, R. J. (1995). DNA solution of hard computational problems. *Science*, 268(5210), 542–545.

Liu, T. Y. and Tsai, W. H. (2007). A new steganographic method for data hiding in Microsoft Word documents by a change tracking technique. *IEEE Transactions on Information Forensics and Security*, 2(1), 24–30.

Lenti, J. (2000). Steganographic methods. *Periodica Polytechnica Electrical Engineering*, 44(3–4), 249–258.

Lewin, D. I. (2002). DNA computing. *Computing in Science and Engineering*, 4(3), 5–8. DOI: 10.1109/5992.998634.

MingXin, L., XueJia, L., GuoZhen, X, and Lei, Q. (2007). Symmetric-key cryptosystem with DNA technology. *Science in China Series F: Information Sciences*, 50(3), 324–333.

Namasudra, S. (2017). An improved attribute-based encryption technique towards the data security in cloud computing. *Concurrency and Computation: Practice and* Exercise. DOI:10.1002/cpe.4364.

Namasudra, S. (2018). Cloud computing: A new era. *Journal of Fundamental and Applied Sciences*, 10(2), 113–135.

Namasudra, S. and Roy, P. (2015). Size based access control model in cloud computing. In *Proceedings of the International Conference on Electrical, Electronics, Signals, Communication and Optimization*, IEEE, Visakhapatnam, India, pp. 1–4.

Namasudra, S. and Roy, P. (2016). Secure and efficient data access control in cloud computing environment: A survey. *Multiagent and Grid System – An International Journal*, 12(2) 69–90.

Namasudra, S. and Roy, P. (2017a). A new secure authentication scheme for cloud computing environment. *Concurrency and Computation: Practice and Exercise*, 29(20). DOI:10.1002/cpe.3864.

Namasudra, S. and Roy, P. (2017b). Time saving protocol for data accessing in cloud computing. *IET Communications*, 11(10), 1558–1565.

Namasudra, S. and Roy, P. (2017c). A new table based protocol for data accessing in cloud computing. *Journal of Information Science and Engineering*, 33(3), 585–609.

Namasudra, S., Roy, P., and Balamurugan, B. (2017a). Cloud computing: Fundamentals and research issues. In *Proceedings of the 2nd International Conference on Recent Trends and Challenges in Computational Models*, IEEE, Tindivanam, India.

Namasudra, S., Roy, P., Balamurugan, B., and Vijayakumar, P. (2017b). Data accessing based on the popularity value for cloud computing. In *Proceedings of the International Conference on Innovations in Information, Embedded and Communications Systems (ICIIECS)*, IEEE, Coimbatore, India, pp. 1–5.

Namasudra, S., Nath, S., and Majumder, A. (2014). Profile based access control model in cloud computing environment. In *Proceedings of the International Conference on Green Computing, Communication and Electrical Engineering*, IEEE, Coimbatore, India, pp. 1–5.

Namasudra, S. and Roy, P. (2018). PpBAC: Popularity based access control model for cloud computing. *Journal of Organizational and End User Computing*, 30(4), 14–31.

National Center for Biotechnology Information. (2018). Retrieved February 17, 2018, from https://www.ncbi.nlm.nih.gov/

Ogihara, M. and Ray, A. (1999). Simulating Boolean circuits on a DNA computer. *Algorithmica*, 25, 239–250.

O'Hare, R. (2016). Microsoft joins the drive to store data in DNA: Firm will buy millions of strands of synthetic genetic code for long-term storage. *MailOnline*, retrieved June 11, 2017, from http://www.dailymail.co.uk/sciencetech/article-3563508/Microsoft-joins-drive-store-data-DNA-Firm-buy-millions-DNA-strands-long-term-storage.html.

Paun, G., Rozenberg G., and Salomaa A. (1998). *DNA Computing: New Computing Paradigms*. Berlin Heidelberg: Springer-Verlag.
Raj, P. and Deka, G. C. (Eds.). (2018). *Blockchain Technology: Platforms, Tools and Use Cases*, 1st edn., Vol. I, Advances in Computers. Cambridge, MA: Academic Press.
Ramsay, T. (2018, January 25). Would you sell your DNA for cryptocurrency? Retrieved February 17, 2018, from https://cryptodaily.co.uk/2018/01/sell-dna-cryptocurrency/.
Reese, A. (2017, December 15). Intel patent describes PoW mechanism that researches DNA. Retrieved February 17, 2018, from https://www.ethnews.com/intel-patent-describes-pow-mechanism-that-researches-dna.
Risca, V. I. (2001). DNA-based steganography. *Cryptologia*, 25(1), 37–49.
Sahai, A. and Waters, B. (2005). Fuzzy identity–based encryption. In Cramer, R. (Ed.), *Advances in Cryptology*, Springer, pp. 457–473.
Sathappan, S. and Pragaladan. R. (2016). DNA based data confidentiality and security in cloud computing. *International Journal of Multidisciplinary Research and Development*. 3(3), 310–312.
Schneier, B. (1996). *Applied Cryptography*, 2nd edn. Hoboken, NJ: John Wiley & Sons.
Sencar, H. T., Ramkumar, M., Akansu, A. N., and Sukerkar A. (2007). Improved utilization of embedding distortion in scalar quantization based data hiding techniques. *Signal Processing*, 87(5), 877–890.
Shamir, A. (1985). Identity-based cryptosystems and signature schemes. In Blakley, G. R. and Chaum, D. (Eds.), *Advances in Cryptology*, Springer, pp. 47–53.
Shyam, M., Kiran, N., and Maheswaran, V. (2007). A novel encryption scheme based on DNA computing. In *Proceedings of the IEEE International Conference*, pp. 1–4.
Satell, G. (2016). Here's how quantum computing will change the world. Retrieved June 11, 2017, from https://www.forbes.com/sites/gregsatell/2016/10/02/heres-how-quantum-computing-will-change-the-world/#675fa7ad6d0a
Gartner. (2017, February 22). Gartner says worldwide public cloud services market to grow 18 percent in 2017. Retrieved June 11, 2017, from http://www.gartner.com/newsroom/id/3616417.
Tsukahara, T. and Nagasawa, H. (2004). Probe-on-carriers for oligonucleotide microarrays (DNA chips). *Science and Technology of Advanced Materials*, 5(3), 359–362.
Voloshynovskiy, S., Pun, T., Fridrich J., Pérez-González, F., and Memon, N. (2003). Security of data hiding technologies. *Signal Processing*, 83(10), 2065–2067.
Weill, K. (2018, January 25). Startup wants to sell your DNA for cryptocurrency. Retrieved February 17, 2018, from https://www.thedailybeast.com/startup-wants-to-sell-your-dna-for-cryptocurrency.
Wikipedia (2018). DNA. Retrieved February 17, 2018, from https://simple.wikipedia.org/wiki/DNA.
Willett, M. (1982). Cryptography old and new. *Computers and Security*, 1(2), 177–186.
Yamamoto, M., Kashiwamura, S., Ohuchi, A., and Furukawa, M. (2008). Largescale DNA memory based on the nested PCR. *Natural Computing, an International Journal*, 7(3), 335–346.

References for Advanced/Further Reading

Chang, W. L. (2012). Fast parallel DNA-based algorithms for molecular computations: Quadratic congruence and factoring integers. *IEEE Transactions on NanoBioscience*, 11(1), 62–69.
Danziger, M. and Henriques, M. A. A. (2012). Computational intelligence applied on cryptography: A brief review. *IEEE Latin America Transactions*, 10(3), 1798–1810.
Namasudra, S., Roy, P, Vijayakumar, P., Audithan, S., and Balamurugan, B. (2017). Time efficient secure DNA based access control model for cloud computing environment. *Future Generation Computer Systems*, 73, 90–105.

2

Implementation of Public Key Cryptography in DNA Cryptography

Alo Sen, Rahul Roy, and Satya Ranjan Dash

CONTENTS

2.1 Introduction

Cryptology is the basis of security for all information. The applications of cryptography are not limited to only mailing, banking, and defense; it has been also implemented in online banking transactions and e-commerce for secure transactions. Security serves to prevent and detect attacks on the authenticated system, data, and network. Cryptology combines two areas (Clelland et al., 1999): *cryptography*, the art of communicating securely through the communication medium; and *cryptanalysis*, the art of deciphering the message by some-one who is not the intended receiver. Both cryptography and steganography are used for providing security, but they are not efficient enough. Many experts say that a security model can be created by adding multiple layers of security to these techniques (Watson and Crick, 1953). A new technology, DNA cryptography, took birth by Adleman in DNA computing and Risca in DNA information project when the most popular cryptographic algorithms, such as Data Encryption Standard (DES) and Adi Shamir, Ron Rivest, and Len Adleman (RSA 768), were also cracked. Then, a new technology was proposed by Adleman and Risca, which is known as DNA cryptography. In DNA computing, DNA

takes on the role of information carrier, and modern biological technologies take on the role of implementation tool. The important features of the DNA molecule stated above can be used for security processes such as signature, authentication, and encryption. No particular specific theory is available for applying DNA molecules in cryptography, but research is still continuing on security-based modern DNA technology.

This chapter proposes a novel method for DNA cryptography using the Diffie Hellman key-sharing technique. The technology/technical terms used in the chapter are explained wherever they appear or in the Key Terminology & Definitions section.

2.2 Literature Survey

DNA cryptography was discovered by Leonard Adleman (1994), who solved the directed Hamiltonian Path problem using DNA computing. It was further developed by (Lipton, 1995), who solved the NP complete problem using DNA computing. Boneh, Dunworth and (Lipton, 1996) confirmed that by using DNA computing, the Data Encryption Standard (DES) cryptographic protocol could be broken. DNA molecules have low energy efficiency, huge parallelism, and extraordinarily large information storage capacity (Cui et al., 2006; Cox, 2001). DNA computing speed can be 1 billion operations per second. DNA molecules have low energy efficiency, huge parallelism, and extraordinarily large information storage capacity (Cui et al., 2006; Cox, 2001). DNA computing speed can be 1 billion operations per second. DNA cryptography (Amos et al, 2002; Xiao et al, 2006) can be added to DNA computing, where DNA will take the role of information carrier and modern biological technologies will take the role of implementation tool. The important features of DNA molecule stated above can be used for security purposes such as signature, authentication, encryption, etc. (Kartalopoulos, 2005). Various methods of DNA cryptography were invented (Cui et al., 2006; Leier et al., 2000) in recent years. No particular specific theory is available to apply DNA molecules in cryptography (Lu et al., 2007; Gehani et al., 2004) but still researches are continuing their researches on security based on modern DNA technology for the successful experimental results. Popular DNA technologies that have been developed in recent years are Polymerase Chain Reaction (PCR), DNA synthesis, and DNA digital coding (Tanaka et al., 2005). According to the Watson-Crick complementary pair, PCR is a very fast DNA amplification technology. With the two primer pairs, namely the forward and reverse primer, the plaintext-encoded sequence can be amplified. DNA hybridization is also a popular technology wherein primary bases (A-T or C-G) get hydrogen bonded together. On the other hand, in traditional cryptography, security is provided via some difficult mathematical calculations. Some traditional efficient existing cryptography methods are AES, DES, and RSA. Hence, traditional cryptography can be combined with DNA cryptography to produce a hybrid cryptography.

Traditional cryptographic algorithms are not suitable for image encryption. So to encrypt images, a new algorithm is required. Image encryption is done in one of two ways, optically or digitally. The digital method, where various encryption algorithms can be used, is used most often; this has more advantages than the optical method, where high-end physical devices are needed, and the digital method is also more efficient than the optical method.

Scrambling the position of the pixel through permutation or by diffusing, through which the pixels in an image can be slightly changed, is not sufficient to save and secure information from statistical attackers. Also, these methods have many limitations in terms of speed and security (Hermassi et al., 2012).

In recent developments, DNA has become the transporter for carrying image information and decrypting images through chaotic maps and enforced images. Fewer computations are required and also the algorithm can be optimized; however, a color-image cryptosystem based on DNA encryption does not work for plaintext (Liu et al., 2012).

The most important issue is how to select the secret key of the algorithm, as most encryption and decryption rules for the plain image are fixed.

An image encryption achieves accuracy through additional operation only after encoding a different phase at pixel level, using the sub-set of DNA complementary rules chaotically. Here the image is first permuted and then encrypted into DNA bases. The adjoining columns of the DNA preset image are supplemented through inter-intra pixel substitution by row addition, called *inter-pixel replacement*, where the DNA base is composed of two bits. To make security stronger, early conditions for chaotic maps are computed from SHA-256 hash of a plain image. Due to there being fewer operations, it is highly efficient (Liao et al., 2018).

In addition, the first secret key that is free of the plaintext, the plaintext-dependent alteration of the key, is also required at the decryption end. Thus, one should choose a different adaptation for different images through a secret path, which can be done by encrypting images using DNA random encoding and self-adaptive permutation–diffusion.

Thus it is not necessary to transfer the secret key privately to the decoder, as the hyper-chaotic Lorenz system generation serves entirely as the secret key.

With the help of self-adaptive permutation–diffusion and DNA random encoding to secure and efficient image encryption, the plain image can be first converted to a DNA sequence, so as to disarrange the bit distribution of the plaintext (Chen et al., 2008), as described in Figure 2.1.

Hyper-chaotic systems have been incorporated into DNA coding to endorse the randomness of the key stream, and they also enhance security (Li et al., 2016).

The secret key and plain image are combined with system parameters and initial conditions where the hash function is used to generate the key streams, which is decomposed with RGB components and translated randomly into three DNA matrices with the help of row-wise and column-wise permutations; the DNA matrix is thus split into three identical parts. A diffusion process is added to the key streams to increase security (Hua et al., 2015).

Low-dimension and high-dimension chaotic maps used in image cryptography have some restrictions (Li et al., 2009): small key space, weak security, high implementation costs, and complex performance analysis (Li, 2016).

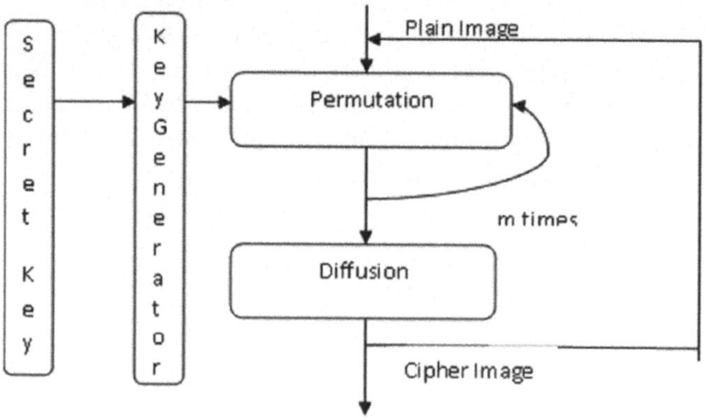

FIGURE 2.1
Permutation-diffusion architecture for image encryption.

Greater storage and enhanced computing capabilities can be achieved through a hybridized model combining a genetic algorithm and DNA sequence, where the correlation among adjacent pixels condensed through the genetic algorithm and next random DNA string resulting in DNA substring after being converted to binary string through XOR operation (Pujari et al., 2018). It is very complicated for intruders to recognize the actual DNA sequence; this can be achieved by appending the DNA sequence from both sides of the original sequence (Majumdar and Sharma, 2014).

Cryptanalytic technology is introduced into DNA cryptography through the DES algorithm in order to improve security and protect information from attackers, but it may lead to high computational complexity (Krishna et al., 2017).

The gap between digital computing and DNA computing lies in the transfer of DNA between two fields, which can reproduce the properties of amino acids. It uses natural RNA codon distribution followed by the conversion of digital data from DNA, which will solve the problem of ambiguity, where more than one codon corresponds to the identical amino acid (Sabry et al., 2017). The key generation process can be done through DNA strands, nucleotides, codons, base pair rules, and mutation (Karimi and Haider, 2017).

We can directly convert to DNA sequence through the frequency of DNA base without converting binary to DNA sequence from static representation; also, security depends on this frequency (Haque and Saha, 2017). Image security can be enhanced through the combination of the trellis algorithm and the DNA sequence, which can be applied both at the time of encryption and decryption (Srividhya and Vino, 2016). Digitally or manually unauthorized access can be prevented through a digital signature by combining fingerprints and iris imaging with the help of DNA cryptography (Mathew and Saranya, 2017).

DNA steganography and DNA-based Advanced Encryption Standard (AES) provides multilayer security for the hidden message. (Sajisha and Mathew, 2017). Data transmitted through the Internet of Things (IoT) and the cloud can be secured using the key generated using the Huffman algorithm and DNA cryptography by using both a symmetric and an asymmetric key (Kumar et al., 2017). By using delayed Hopfield NN after generating key in first level through DNA cryptographic model can be used to encrypt the text. Based on DNA sequence, we can have Network Intrusion Detection System (NIDS), which can discover patterns, followed by converting the network traffic data into the form of DNA sequence after that only classification can be done. But the sequence generation mainly depends on the DNA encoding techniques used (Rashid et al., 2017). By combining AES, DNA cryptography, and audio steganography, video piracy can be detected (Ajaykumar et al., 2017).

DNA-encoded diffused images under the control of a key stream can be confused by the introduction of DNA deletions and DNA insertions. The encoded image can be obtained after decoding the confused image (Hu et al., 2017). An Exclusive OR (XOR) operation, performed between the secret image and the key matrix, is used to create multiple sharing obtained from Boolean-based (n,n) multi-secret-image sharing scheme to protect the digital media information (Eswaran and Shankar, 2017).

2.3 Preliminaries

2.3.1 Symmetric Key Cryptosystem

Symmetric encryption, a conventional or one-key encryption method, was the only type of encryption before the development of public-key encryption. In symmetric encryption,

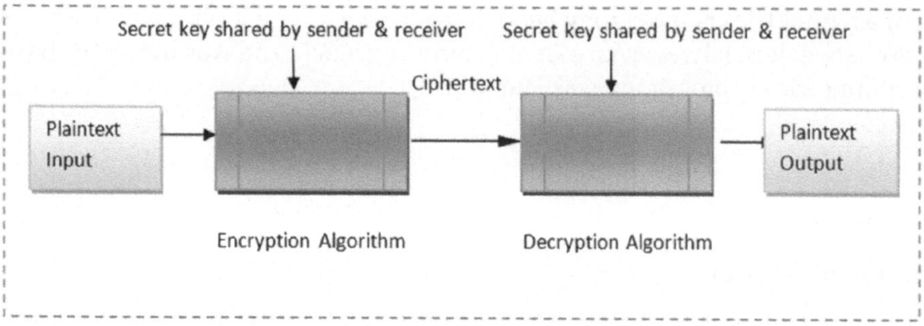

FIGURE 2.2
The symmetric encryption scheme.

a single key is used to both encrypt and decrypt a message or plaintext. The symmetric encryption scheme has total of five components (Figure 2.2):

1. *Plaintext*: Plaintext is the original message or data that is to be encrypted.
2. *Encryption algorithm*: The encryption algorithm performs different operations on the plaintext to encrypt the message in secure way.
3. *Secret key*: The secret key is a very important aspect of the encryption algorithm. This is the key through which encryption and decryption can be performed on the plaintext.
4. *Ciphertext*: The encrypted message is known as ciphertext. It depends entirely on the plaintext and the key. The ciphertext is generally an encrypted form of plaintext that needs to be decrypted on the receiver side.
5. *Decryption Algorithm*: The decryption algorithm decrypts the message, working in the reverse order of the encryption algorithm. It takes the ciphertext and the secret key as input and generates the original plaintext as an output sent by the sender.

There are two problems with symmetric key cryptography:

1. *Distribution of the symmetric key to be shared by sender and receiver* – If there are n people communicating with each other, $n \times (n-1)/2$ symmetric keys need to be distributed between them.
2. The concept of *digital signature* is important if the sender and receiver do not really trust one another, and the sender later claims that she/he did not send the receiver any message or the receiver claims that she/he did not receive any message from the sender.

2.3.2 Requirements for a Symmetric Key Cryptosystem

Two requirements need to be considered for symmetric key cryptosystems:

1. The secret key
2. A secure single channel through which sender and receiver can communicate

2.3.3 Diffie Hellman Key-Sharing Technique

The Diffie Hellman key-sharing method involves sharing a public key between the sender and receiver, through which they can compute a secret key by having each other's public key. But even though it uses the same underlying principles as public key cryptography,

this is not asymmetric cryptography because nothing is ever encrypted or decrypted during the exchange. It is, however, an essential building block, and was in fact the basis upon which asymmetric cryptography was later built.

2.4 Proposed Model

In the proposed approach (Figure 2.3.) a shared secret key–based DNA cryptosystem is proposed. The symmetric key cryptosystem has been developed with different DNA technologies like a DNA encoding scheme, DNA hybridization, and DNA primers. The encryption process is done with DNA hybridization technology. This proposed algorithm is based on a symmetric key cryptography system where a shared secret key is used for encryption as well as decryption of the message. The Diffie Hellman key-exchange technique is a very famous and highly appropriate technique for computing the cryptographic shared secret key. Here, researchers have used this key-sharing technique, where the original shared secret key is not actually transmitted over the channel. The two parties, sender as well as receiver, have to agree on two values. For that purpose, researchers have used the publicly available National Council of Biotechnological Information (NCBI) database to choose one agreed-upon value and primers. The sender must send primers, an organism name with the NCBI database and the type of genomic data, and the two agreed-upon values for computation of the shared secret key to the receiver.

In Figure 2.3, the plaintext is being converted into DNA coded output by using the method shown in Table 2.1. Then the DNA coded output is placed between the forward and end primer. Then the total bits are converted in a corresponding binary sequence using Table 2.2. The DNA equivalent sequence of the shared secret key is used for DNA hybridization with the binary sequence to produce the ciphertext. Then the ciphertext is transferred through a communication channel to reach the receiver. After receiving the

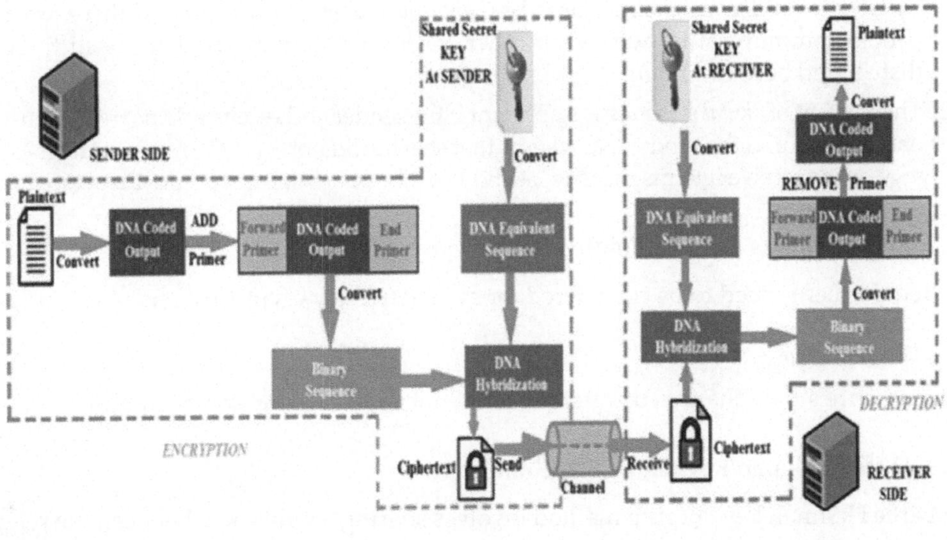

FIGURE 2.3
Proposed model.

TABLE 2.1

Translation Table for Alphabets, Digits, and Punctuation Marks in DNA Bases

A=CGA	H=CGC	O=GGC	V=CCT	2=TAG	9=GCG
B=CCA	I=ATG	P=GGA	W=CCG	3=GCA	.=ATA
C=GTT	J=AGT	Q=AAC	X=CTA	4=GAG	,=TCG
D=TTG	K=AAG	R=TCA	Y=AAA	5=AGA	.=GAT
E=GGT	L=TGC	S=ACG	Z=AAT	6=GGG	:=GCT
F=ACT	M=TCC	T=TTC	0=TTA	7=ACA	;=ATT
G=TTT	N=TCT	U=CTG	1=ACC	8=AGG	-=ATC

TABLE 2.2

Binary DNA Coded Scheme

Nucleotide	A	C	G	T
Code	00	01	10	11

ciphertext, the receiver follows a reverse procedure of DNA hybridization with the shared secret key to get the binary sequence. Then, the binary sequence is converted by adding forward and end primers with the DNA coded output. The primers are removed in a reverse process to retrieve the DNA coded output, and finally, it is converted into plaintext on the receiver end.

The proposed algorithm is fragmented into encryption, decryption, and key-generation parts respectively in the following subsections 2.4.1 through 2.4.3. Researchers have chosen the nucleotide sequence of *Canis lupus familiaris* to determine the primers and one agreed-upon value for the Diffie Hellman key-sharing technique.

2.4.1 Key Generation Procedure

Researchers have used the Diffie Hellman key-sharing technique to produce the shared secret key. Following the Diffie Hellman technique, the sender and receiver first agree upon two values (n = a large prime number, g = generator of n). Then, they decide their private key and compute their public key. After that, they exchange their public key with each other through the open communication channel. Finally, they calculate the shared secret key that will be used to encrypt the message one the sender side as well as to decrypt the message on the receiver side. Figure 2.4 illustrates the shared secret-key generation procedure.

Sender and receiver have their own private keys, namely X_A and X_B respectively. Then sender and receiver calculate $\mathrm{Y_A}$ and $\mathrm{Y_B}$ respectively and exchange the values with each other. Finally, on both sides, the shared secret key is calculated and converted into a corresponding ATCG sequence, which is used for encryption and decryption of messages at both ends.

2.4.2 Encryption Process

Encryption of the plaintext to ciphertext involves the following steps:

1. The plaintext is transferred into DNA coded output by following DNA coded set that was proposed by (Clelland et al., 1999), where DNA coding can be expressed only for capital letters, digits, and punctuation marks (cf. Table 2.1).

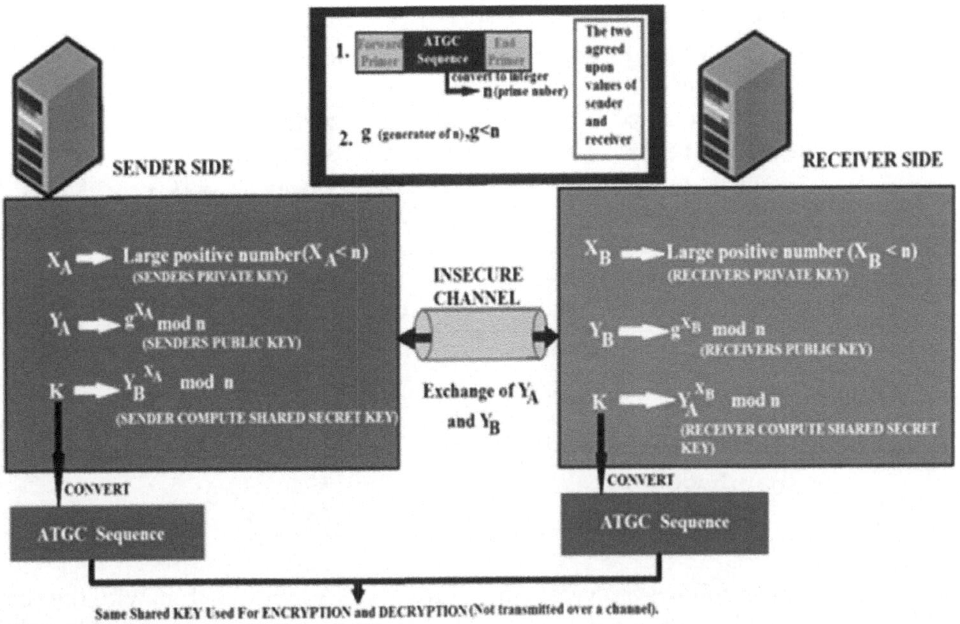

FIGURE 2.4
Procedure for creating a shared secret key in the proposed system using the Diffie Hellman key-sharing technique.

Forward Primer	DNA coded output	End Primer

FIGURE 2.5
Structure of the DNA sequence with primers.

The DNA coded set is made up as follows:

2. Extra bits (i.e., forward primers and end primers) have to be attached with the above DNA coded output (cf. Figure 2.5).
3. The whole DNA sequence is converted into a binary equivalent sequence with the following binary DNA coded scheme (c.f Table 2.2).
4. In the process of encrypting the plaintext, the DNA hybridization is performed only for the binary '1' in the binary equivalent sequence. If the binary bit is found to be '0', no operation takes place. Wherever the binary digit is '1', then the corresponding 10 bases of the DNA sequence of the shared secret key are complemented to form an oligonucleotide sequence, and thus the plaintext is finally converted to ciphertext. The digits are compared with the DNA sequence in reverse order.
5. The ciphertext is then transferred through the open communication channel.

2.4.3 Decryption Process

Decryption involves the following steps:

1. DNA hybridization take place between the received ciphertext and the DNA equivalent sequence of the shared secret key in the following way:

In the binary sequence, '1' will be considered if every 10 bases of the ciphertext are equal to the complementary form of the 10 bases of the shared secret key from the reverse order; otherwise it will be considered as '0'.

2. Then the generated binary sequence is converted to the DNA coded output with forward and end primer using the abovementioned binary coded scheme.
3. Finally, the two primers are removed from the DNA coded output and then the DNA coded output is converted to plaintext, which is the original message sent by the sender.

2.4.4 Result and Discussion

A numerical experiment has been done to explain the working principle of the proposed algorithm. Researchers have considered the plaintext 'CAT'. First, one random nucleotide sequence is chosen from the NCBI database of the organism DOG. The scientific name of the organism is *Canis lupus familiaris*. Then, the following nucleotide sequence is chosen to decide the primers and value of *n* (one of the agreed-upon values for sender and receiver) for the Deffie Hellman key-sharing technique:

GTCCACATCATTGGAATGCCATTTCTTATTCATCAGTGGGCCCGGGGAG-GAGAGTGGGTGTTTGGGGGGCCCCTCTGCACCATTATCACATCCCTGGATACCT-GCAACCAGTTTGCCTGTAGTGCCATCATGACTGTGATGAGTATAGACAG-GTACTTGGCTCTCGTCCAACCATTTCGACTTACAAGTTGGAGAACGAG-GTACAAGACCATCCGCATCAATTTGGGCCTTTGGGCAGCTTCCTTCAT TCTGGCGCTGCCTGTCTGGGTCTACTCGAAGGTCATCAAATTTAAAGACGGCGTG-GAGAGTTGTGCTTTTGATTTAACATCCCCTGACGATGTACTCCGGTATACACTT-TATTTGACGATAACAACTTTTTTTTTCCCTTTGCCTTTGATTTTGGTGTGCTATATTT-TAATTTTATGCTATACTTGGGAGATGTATCAACAGAATAAAGATGCAAGATGTTA-CAATCCCAGTGTTCCAAAAGAGAGAGTGATGAAGCTGACAAAGATGGTGCTG-GTGCTGGTGGCGGTCTTTATCCTAAGTGCTGCCCCCTACCACGTGATACAACTGGT-GAACTTAAAGATGCAGCAGCCC.

GTCCACATCA: The forward primer, GCAGCAGCCC: The end primer, *n*: The nucleotide sequence between forward and end primer (total 580 bits length). At first, the plaintext is converted to the following DNA coded output: GTTCGATTC. Then, the forward primer is added to the header of the DNA coded output and the end primer is added to the end of the DNA coded output. The converted DNA sequence with primers:

GTCCACATCAGTTCGATTCGCAGCAGCCC.

Now, the binary equivalent of the DNA sequence with primer is:

1101101000100001100011010110110001011011100011100011101010 (Total 58 bits length).

For generating a shared secret key, the Diffie Hellman key-sharing technique is applied. Here, *n* is chosen as nucleotide sequence. This *n* must then be converted into an integer value, which will also be a prime number. So, the converted nearest primer value of the nucleotide sequence is 42179. Let the value of g = 983, generator of *n*; the sender and receiver will both agree upon those two values. For simplicity's sake, the private keys of the sender and receiver are considered as 1 and 70, respectively. Now, they will calculate their public key using the Diffie Hellman public-key creation formula. After computation, the public key of sender = 983.0, which will be transferred to the receiver; and the public key of receiver = 30748.0, which will be transferred to the sender through an insecure channel. Then they will calculate the shared secret key: For both receiver and sender, the secret key is 30748.0. Now, the shared secret key will be converted to the

equivalent DNA sequence: GCATTAACAGAGAGG. But, to perform DNA hybridization, it is required that the length of equivalent DNA sequence of shared secret key be 580 bits, as the length of the DNA coded output with primer = 58. So, the length of the equivalent DNA sequence of the shared secret key is made equal by concatenating the same sequence until the length is 580. The resultant equivalent DNA sequence of the shared secret key 580 bits in length is:

GCATTAACAGAGAGGGCATTAACAG.

Then, DNA hybridization is done with the DNA sequence of the shared secret key and the DNA coded output with primers. The DNA form of encrypted message (ciphertext), which is also 580 bits in length, is:

CGTAATTGTCTTGTCTCTCCAGAGGGCATTCGTAATTGTCTTGTCTCTCCAGAGGGCATTCGTAATTGTCAACAGAGAGGAGAGGGCATTGCATTAACAGTTGTCTCTCCAGAGGGCATTGCATTAACAGAACAGAGAGGAGAGGGCATTCGTAATTGTCTTGTCTCTCCAGAGGGCATTGCATTAACAGAACAGAGAGGTCTCCCGTAACGTAATTGTCAACAGAGAGGTCTCCCGTAAGCATTAACAGTTGTCTCTCCTCTCCCGTAAGCATTAACAGTTGTCTCTCCTCTCCCGTAAGCATTAACAGAACAGAGAGGAGAGGGCATTCGTAATTGTCAACAGAGAGGTCTCCCGTAACGTAATTGTCAACAGAGAGGTCTCCCGTAACGTAATTGTCTTGTCTCTCCAGAGGGCATTGCATTAACAGAACAGAGAGGTCTCCCGTAACGTAATTGTCTTGTCTCTCCAGAGGGCATTGCATTAACAGAACAGAGAGGTCTCCCGTAACGTAATTGTCTTGTCTCTCCAGAGGGCATTCGTAATTGTCAACAGAGAGGTCTCCCGTAAGCATTAACAG.

The above ciphertext is transferred to the receiver through a channel. Now, after receiving the above ciphertext, the receiver will also do the DNA hybridization of the ciphertext with the shared secret key. The corresponding result of the sequence, which is also 580 bits in length, is:

GCATTAACAGAACAGAGAGGTCTCCCGTAAGCATTAACAGAACAGAGAGGTCTCCCGTAAGCATTAACAGTTGTCTCTCCTCTCCCGTAACGTAATTGTCAACAGAGAGGTCTCCCGTAACGTAATTGTCTTGTCTCTCCTCTCCCGTAAGCATTAACAGAACAGAGAGGTCTCCCGTAACGTAATTGTCTTGTCTCTCCAGAGGGCATTGCATTAACAGTTGTCTCTCCAGAGGGCATTCGTAATTGTCAACAGAGAGGAGAGGGCATTCGTAATTGTCAACAGAGAGGAGAGGGCATTCGTAATTGTCTTGTCTCTCCTCTCCCGTAAGCATTAACAGTTGTCTCTCCAGAGGGCATTGCATTAACAGTTGTCTCTCCAGAGGGCATTGCATTAACAGAACAGAGAGGTCTCCCGTAACGTAATTGTCTTGTCTCTCCAGAGGGCATTGCATTAACAGAACAGAGAGGTCTCCCGTAACGTAATTGTCTTGTCTCTCCAGAGGGCATTGCATTAACAGAACAGAGAGGTCTCCCGTAAGCATTAACAGTTGTCTCTCCAGAGGGCATTCGTAATTGTC.

The receiver will convert the DNA sequence to a binary sequence 58 bits in length using the above table: 11011010001000001100011010110110001011011100011100011101010. After this

FIGURE 2.6
Output of the experimental results of the proposed system.

binary sequence is converted to a DNA sequence, the result will be the following DNA coded output with primers:

GTCCACATCAGTTCGATTCGCAGCAGCCC. Then, the receiver will have to remove the forward as well as the end primer to get the original DNA coded output. So, the original DNA coded output will be GTTCGATTC. Then, the DNA coded output is converted to the original plaintext by using the above table, which results in 'CAT'. Figure 2.6 depicts the experimental result implemented by Java.

2.5 Security Analysis

In the modern era of DNA computing, various encryption schemes have been proposed using different DNA technologies like DNA digital coding, PCR primers, DNA synthesis, DNA hybridization, etc. Among these various encryption schemes, a popular scheme was proposed by (Cui et al., 2007, 2008), an asymmetric cryptosystem wherein a plaintext is converted to a DNA template where primers were used as key to encode and decode the message. But DNA encryption with PCR may not be always safe because:

- The relationship between the original plaintext and the converted DNA sequence is one-to-one; thus, the statistical properties of plaintext are in built into the ciphertext.

So, it is possible for an intruder to attack the cryptosystem through a statistical attack, which leads to a security risk, breaking the first level of security in the system.

Another author (Shreyas Chavan, 2013) has proposed an encryption scheme using DNA hybridization with a one-time-pad scheme. In this scheme, random ssDNA and random OTP keys are generated and are hybridized to create the ciphertext. Though this algorithm is scalable, there is a chance that the intruder could break through the security due to the following reasons:

1. The cryptosystem is based on a traditional binary DNA encoding scheme to convert the plaintext into DNA sequence.
2. There is a chance of a man-in-the-middle attack, where an intruder can get access to the generated random OTP key. Also, with respect to the security aspects, reusing the same random key for multiple functions in the cryptosystem can open up security holes, which could lead to a protocol attack in which an intruder implements a fresh protocol using the same key as a targeted protocol that is very strong individually (Kelsey et al., 1998).

2.5.1 Security Features

In the modern era, every possible type of security risk should be considered effectively. It should be considered that an intruder knows the basic DNA technology that the designers can use, and they may also have extensive knowledge as well as modern experimental devices that they can use to follow the proposed encryption method. So, whatever the advanced DNA technology used, it still carries a great security risk. One solution can be found in key generation as well as the key-sharing procedure.

1. In this proposed symmetric cryptosystem, the actual key for encryption and decryption is not transmitted over the channel. Rather, this research used the Diffie Hellman key-sharing technique to compute the shared secret key on the both the sender's and receiver's sides by only exchanging each other's public key.
2. The size of the key is one of the most important factors for the security of the cryptosystem. An efficient cryptosystem should have a key that is big enough to ensure its safety. The traditional encryption PCR method of encrypted DNA sequences runs into a problem if there is not enough key space for a secure cryptosystem. In this proposed method, the Diffie Hellman key-sharing technique will solve this problem by having a large prime number that can make the sender's and receiver's private key bigger.

The security of the proposed cryptosystem has two levels: The first level is biological information security and the complexity of the biologically inspired difficult (DNA hybridization) problems. The extra bits (forward and end primers) in the original DNA sequence create a fake DNA sequence; this is hybridized with the shared secret key to form the ciphertext, making the plaintext less vulnerable to attackers. It is more difficult for attackers to predict the main ciphertext as it contains extra information in the form of the forward primer and end primer. The intruders have to know both the primers, otherwise the DNA hybridization will fail to give the correct ciphertext. If they get access to the primers, then the security of the cryptosystem will remain intact. The second level of

security is the modular arithmetic problem of the Diffie Hellman key-sharing technique. Without getting the receiver's private key, it is not possible for an intruder to calculate the shared secret key; moreover, without the receiver's key, the intruder would have to do a large and difficult calculation to break the security of the proposed cryptosystem. Thus this cryptosystem is highly secure because of the difficulties in biological issues and computation of shared secret-key difficulties added a double security level for the system.

2.6 Performance Study of Diffie Hellman Key-Sharing Technique

DNA cryptography can offer high levels of security using the above proposed approach, but it has vulnerabilities, like the possibility of a man-in-the-middle attack in areas of the Diffie Hellman key-exchange technique. It is always not that only one man will attack in the key-sharing technique rather than N party man in middle attack is also possible where every man in the middle will have access to its neighbors' public components and will generate their own keys similar to his/her neighbor. Now, researchers are focusing on some existing work to prevent man in middle attack for Diffie Hellman.

The Station-to-Station (STS) protocol was proposed (Diffie et al., 1992) to prevent man-in-the-middle attacks. Secure Sockets Layer (SSL) (Frier and Kocher, 1996) involves negotiating and establishing secure connections and securing the data transmission. The SSL handshake uses certificates and PKI (Younglove, 2001) for mutual authentication and key exchange. PKI binds public keys with particular user identities by means of a certificate authority (CA).

An algorithm to prevent a man-in-the-middle attack is shown in Figure 2.7, illustrating the respective algorithm in terms of a flow chart (Table 2.3).

Another method for preventing a man-in-the-middle attack (Khader and Lai, 2015) is shown in the corresponding diagram in Figure 2.8.

A novel method to illustrate *N* party (Devi and Makani, 2015) man-in-the-middle attack in the Diffie Hellman key-sharing technique, is shown in Figure 2.9, which depicts the block diagram of *N* party man-in-the-middle attack.

2.7 Conclusion

The proposed algorithm represents a new approach in symmetric key cryptography, where the shared secret key is generated using the Diffie Hellman key-sharing technique. The generation of a shared secret key is the highest level of security against intruders breaking the cryptosystem. Other DNA technologies like the DNA coding scheme (cf. Table 2.1) and DNA hybridization are used with this cryptosystem. The algorithm is more reliable and scalable, and it also gives robust performance. Adding extra information to the ciphertext makes the algorithm more secure. However, the manipulation of DNA sequences takes a lot of time compared to that required by many very efficient algorithms of traditional cryptography such as DES, RSA, etc. (Beaver, 1995; Brun, 2007). It can face problems due to the huge computing time required and its high computational complexity.

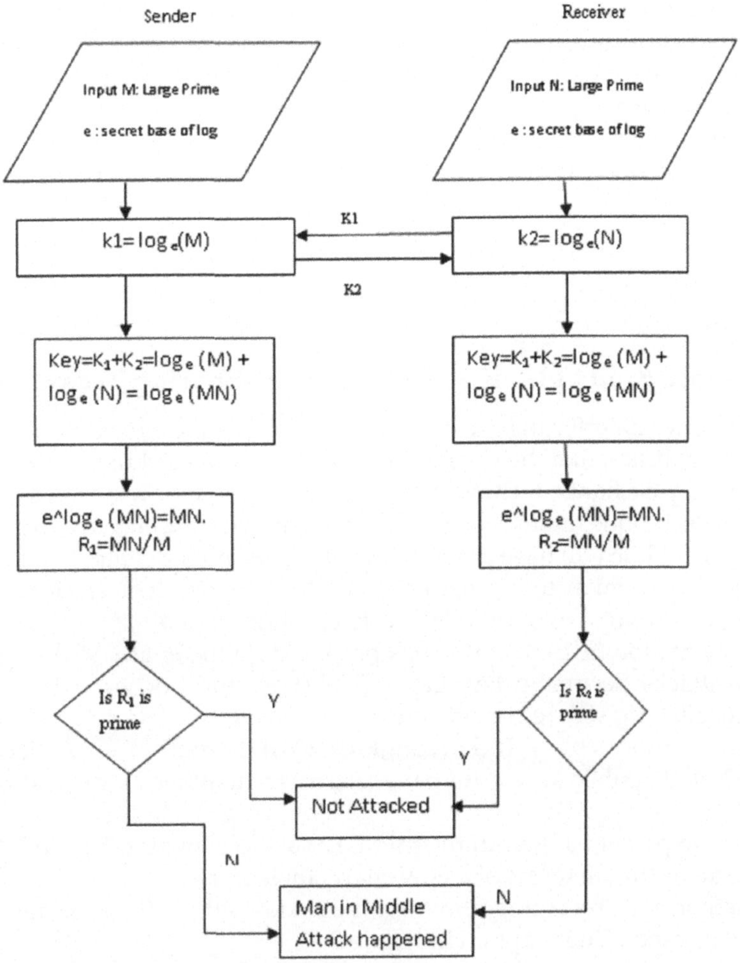

FIGURE 2.7
Existing algorithm to prevent man-in-the-middle attacks.

TABLE 2.3
Comparison Among Existing Models to Prevent Man-in-the-Middle Attacks

Serial No	Authors	Methods	Advantages	Disadvantages
1	Biswas and Basuli (2012)	Logarithm method has been adopted for the mathematical calculation	Reduces the overhead of data or key exchange between nodes	Man-in-the-middle attack is not fully dissolved because the base selected by the hacker may be same as chosen by B and K
2	Khader and Lai (2015)	Users have usernames and passwords, and these will be used to connect them with their systems	Distribute the keys in secure manner	The authors have used a third-party server who is controlling the private keys with an admin key. Here, maintaining the server is expensive and there will be an overhead if the server breaks down

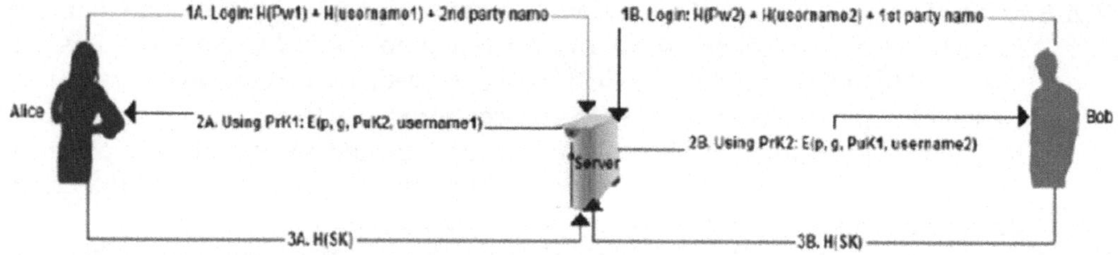

FIGURE 2.8
Alice and Bob communicate with the server to obtain the shared key. (From Khader, A. S., and Lai, D. Preventing man-in-the-middle attack in Diffie Hellman key-exchange protocol. In *2015 22nd International Conference on Telecommunications (ICT)*, pp. 204–208, IEEE, 2015).

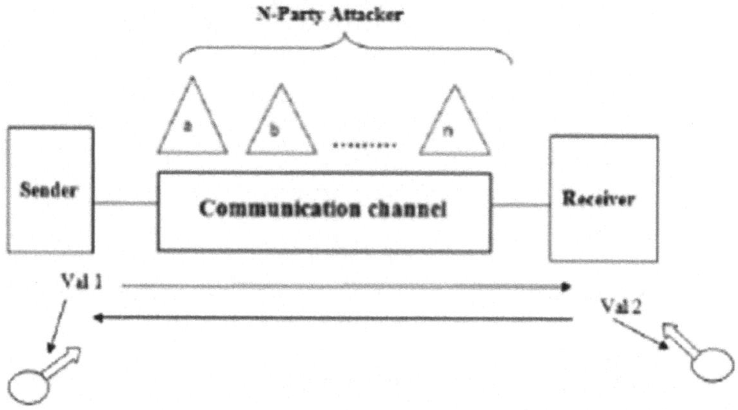

FIGURE 2.9
Block diagram of N party man-in-the-middle attack. (From Devi, S. and Makani, R. *International Journal of Computer Science and Information Technologies*, 6 (5), 4281–4285, 2015.)

In future, the proposed cryptosystem can be used for the real-time security of the distributed networking system. It can be enhanced with much more advanced topics like realization of many security technologies available for encryption, signature, steganography, and authentication. The hardware implementation can be easy for this simple cryptosystem and it can be very efficient. Multiple rounds can also be implemented easily using advanced hardware technology. This cryptosystem is thus suitable for both hardware as well as software implementation.

Key Terminology and Definitions

Public Key Cryptography: Public key cryptography, or asymmetrical cryptography, is any cryptographic system that uses pairs of keys:

- Public keys, which may be disseminated widely
- Private keys, which are known only to the owner. Unlike symmetric key algorithms that rely on one key to both encrypt and decrypt, each key performs a unique function.

DNA hybridization: DNA hybridization refers to a molecular biology technique that measures the degree of genetic similarity between pools of DNA sequences. It is usually used to determine the genetic distance between two organisms.

Encryption and Decryption: Encryption is the process of translating plaintext data (plaintext) into something that appears to be random and meaningless (ciphertext). Decryption is the process of converting ciphertext back to plaintext. To encrypt more than a small amount of data, symmetric encryption is used.

Diffie Hellman key exchange: Diffie Hellman key exchange, also called *exponential* key exchange, is a method of digital encryption that uses numbers raised to specific powers to produce decryption keys on the basis of components that are never directly transmitted, making the task of a would-be code breaker mathematically overwhelming.

Authentication: Authentication refers to a group of processes where confidence in user identities is established and presented via electronic methods to an information system. It is also referred to as *e-Authentication.*

References

Adleman, L. M. (1994). Molecular computation of solutions to combinatorial problems. *Science-AAAS-Weekly Paper Edition*, 266(5187), 1021–1023.

Ajaykumar, A., Juvin Varghese, A. C., and Kodakara, S. C. E. T. (2017). Video piracy detection based on audio steganography, AES and DNA cryptography. *International Journal of Engineering Science*, 5487.

Amos, M., Păun, G., Rozenberg, G., and Salomaa, A. (2002). Topics in the theory of DNA computing. *Theoretical Computer Science*, 287(1), 3–38.

Beaver, D. (1995). Factoring: The DNA solution. In *Advances in Cryptology—ASIACRYPT'94* (pp. 419–423). Springer: Berlin Heidelberg.

Biswas, B. and Basuli, K. (2012). A novel process for key exchange avoiding man-in-middle attack. *International Journal of Advancements in Research and Technology*, 1(4), 75–79.

Brun, Y. (2007). Arithmetic computation in the tile assembly model: Addition and multiplication. *Theoretical Computer Science*, 378(1), 17–31.

Chen, J., Zhu, Z. L., Zhang, L. B., Zhang, Y., and Yang, B. Q. (2018). Exploiting self-adaptive permutation-diffusion and DNA random encoding for secure and efficient image encryption. *Signal Processing*, 142, 340–353.

Clelland, C. T., Risca, V., and Bancroft, C. (1999). Hiding messages in DNA microdots. *Nature*, 399(6736), 533–534.

Cox, J. P. (2001). Long-term data storage in DNA. *TRENDS in Biotechnology*, 19(7), 247–250.

Cui, G. Z., Liu, Y., and Zhang, X. (2006). New direction of data storage: DNA molecular storage technology. *Computer Engineering and Applications*, 42(26), 29–32.

Cui, G., Qin, L., Wang, Y., and Zhang, X. (2007). Information security technology based on DNA computing. In *2007 IEEE International Workshop on Anti-Counterfeiting, Security, Identification*, pp. 288–291. IEEE.

Cui, G., Qin, L., Wang, Y., and Zhang, X. (2008). An encryption scheme using DNA technology. In *3rd International Conference on Bio-Inspired Computing: Theories and Applications, 2008. BICTA 2008*, pp. 37–42. IEEE.

Devi, S., and Makani, R. (2015). Generation of n-party man-in-middle attack for Diffie-Hellman key exchange protocol: A review. *International Journal of Computer Science and Information Technologies, Vol. 6* (5), 4281–4285.

Diffie, W., Van Oorschot, P.C., Wiener, M. J. (1992) Authentication and authenticated key exchanges. *Des. Codes Cryptography* 2(2), 107–125.

Eswaran, P. and Shankar, K. (2017), Multi secret image sharing scheme based on DNA cryptography with XOR. *International Journal of Pure and Applied Mathematics*, 118(7), 393–398.

Frier, A. and Kocher, K. P., (1996). The secure socket layer. Technical report, Netscape Communications Corp.

Gehani, A., LaBean, T., and Reif, J. (2004). DNA-based cryptography. In *Aspects of Molecular Computing* (pp. 167–188). Springer: Berlin Heidelberg.

Hermassi, H., Rhouma, R., and Belghith, S. (2012). Security analysis of image cryptosystems only or partially based on a chaotic permutation, *Journal of Systems and Software* 85(9), 2133–2144.

Haque, R. and Saha, R. (2017). A novel rolling based DNA cryptography. *Journal of Bioinformatics and Genomics*, 1(3).

Hu, T., Liu, Y., Gong, L. H., Guo, S. F., and Yuan, H. M. (2017). Chaotic image cryptosystem using DNA deletion and DNA insertion. *Signal Processing*, 134, 234–243.

Hua, Z., Zhou, Y., Pun, C. M., and Chen, C. P. (2015). 2D sine logistic modulation map for image encryption. *Information Sciences*, 297, 80–94.

Karimi, M. and Haider, W. (2017). Cryptography using DNA nucleotides. *International Journal of Computer Applications*, 168, 16–18.

Kartalopoulos, S. V. (2005). DNA-inspired cryptographic method in optical communications, authentication and data mimicking. In *Military Communications Conference, 2005. MILCOM 2005*, pp. 774–779. IEEE.

Kelsey, J., Schneier, B., and Wagner, D. (1998). Protocol interactions and the chosen protocol attack. In *Security Protocols* (pp. 91–104). Springer: Berlin Heidelberg.

Khader, A. S. and Lai, D. (2015). Preventing man-in-the-middle attack in Diffie-Hellman key exchange protocol. In *2015 22nd International Conference on Telecommunications (ICT)*, pp. 204–208. IEEE.

Krishna, B. M., Khan, H., Madhumati, G., Kumar, K. P., Tejaswini, G., Srikanth, M., and Ravali, P. (2017). FPGA implementation of DES algorithm using DNA cryptography. *Journal of Theoretical and Applied Information Technology*, 95(10).

Kumar, N. H., Patil, R. M., Deepak, G., and Murthy, B. M. (2017, March). A novel approach for securing data in IoT cloud using DNA cryptography and Huffman coding algorithm. In *2017 International Conference on Innovations in Information, Embedded and Communication Systems (ICIIECS)*, pp. 1–4. IEEE.

Liu, L., Zhang, Q., and Wei, X.A., (2012). RGB image encryption algorithm based on DNA encoding and chaos map. *Computers and Electrical Engineering*, 38, 1240–1248.

Leier, A., Richter, C., Banzhaf, W., and Rauhe, H. (2000). Cryptography with DNA binary strands. *BioSystems*, 57(1), 13–22.

Li, C. (2016). Cracking a hierarchical chaotic image encryption algorithm based on permutation. *Signal Processing*, 118, 203–210.

Li, C., Li, S., Chen, G., and Halang, W. A. (2009). Cryptanalysis of an image encryption scheme based on a compound chaotic sequence. *Image and Vision Computing*, 27(8), 1035–1039.

Li, X., Wang, L., Yan, Y., and Liu, P. (2016). An improvement color image encryption algorithm based on DNA operations and real and complex chaotic systems. *Optik-International Journal for Light and Electron Optics*, 127(5), 2558–2565.

Liao, X., Hahsmi, M. A., and Haider, R. (2018). An efficient mixed inter-intra pixels substitution at 2bits-level for image encryption technique using DNA and chaos. *Optik-International Journal for Light and Electron Optics*, 153, 117–134.

Lipton, R. J. (1995). DNA solution of hard computational problems. *Science*, 268(5210), 542–545.

Lipton, R. J. (1996). Breaking DBS using a molecular computer Dan Boneh Christopher Dimworth. *DNA Based Computers*, 27, 37.

Lu, M., Lai, X., Xiao, G., and Qin, L. (2007). Symmetric-key cryptosystem with DNA technology. *Science in China Series F: Information Sciences*, 50(3), 324–333.

Majumdar, A. and Sharma, M. (2014). A new approach towards information security based on DNA cryptography, 4(4), 59–68.

Mathew, S. and Saranya, G. (2017). Advanced biometric home security system using digital signature and DNA cryptography. In *2017 International Conference on Innovations in Green Energy and Healthcare Technologies (IGEHT)*, pp. 1–4. IEEE.

Pujari, S. K., Bhattacharjee, G., and Bhoi, S. (2018). A hybridized model for image encryption through genetic algorithm and DNA sequence. *Procedia Computer Science*, 125, 165–171.

Rashid, O. F., Othman, Z. A., and Zainudin, S. (2017). A novel DNA sequence approach for network intrusion detection system based on cryptography encoding method. *International Journal on Advanced Science, Engineering and Information Technology*, 7(1), 183–189.

Sabry, M., Hashem, M., and Nazmy, T. (2017). Digital encoding to the form of amino acids for DNA cryptography and biological simulation. *International Journal of Computer Applications*, 165(10).

Sajisha, K. S., and Mathew, S. (2017). An encryption based on DNA cryptography and steganography. In *2017 International Conference of Electronics, Communication and Aerospace Technology (ICECA)*, Vol. 2, pp. 162–167. IEEE.

Shreyas Chavan (2013). DNA cryptography based on DNA hybridization and one time pad scheme. *International Journal of Engineering Research and Technology (IJERT)*, 2, 10.

Srividhya, N. and Vino, T. (2016, March). Genome based highly secured image using DNA cryptography and trellis algorithm. In *International Conference on Wireless Communications, Signal Processing and Networking (WiSPNET)*, pp. 1658–1662. IEEE.

Tanaka, K., Okamoto, A., and Saito, I. (2005). Public-key system using DNA as a one-way function for key distribution. *Biosystems*, 81(1), 25–29.

Watson, J. D. and Crick, F. H. C. (1953). A structure for deoxyribose nucleic acid. *Nature*, 421(6921), 397–398.

Xiao, G., Lu, M., Qin, L., and Lai, X. (2006). New field of cryptography: DNA cryptography. *Chinese Science Bulletin*, 51(12), 1413–1420.

Younglove, R. (2001). Public key infrastructure. How it works. *Computing and Control Engineering Journal*, 12, 99–102.

References for Advanced/Further Reading

Deka, G. C. (Ed.). (2014). *Handbook of Research on Securing Cloud-Based Databases with Biometric Applications*. IGI Global.

Muthulakshmi, B. and Venkatesulu, M. (2018). *Data Partitioning in Cloud Storage Using DESD Crypto Technique*. IJCSIS.

Sarkar, A. (2018). Genetic Key Guided Neural Deep Learning Based Encryption for online wireless communication (GKNDLE). *International Journal of Applied Engineering Research*, 13(6), 3631–3637.

Tomar, G. S., Chaudhari, N. S., Bhadoria, R. S., and Deka, G. C. (Eds.). (2016). *The Human Element of Big Data: Issues, Analytics, and Performance*. CRC Press: Boca Raton, FL.

Vikas, B., Akshay, A. K., Thanneeru, S. P. M., Raghuram, U. M. V., and Bhargav, K. S. (2018). A novel DNA-and PI-based key generating encryption algorithm. In *Information Systems Design and Intelligent Applications*, pp. 945–954. Springer: Singapore.

3

Taxonomy of DNA-Based Security Models

Suyel Namasudra

CONTENTS

3.1 Introduction

In the 21st century's digital era, data or information is everywhere. Data transactions and data communication have been revolutionized by the internet. Information or data security is very important in all domains, from the level of the individual user to the national level. The cryptographic approach is the main kernel of all data or information security technology. *Cryptography* is the technique of secret writing (Willett, 1982; Lin, 1998). In its earlier stages, cryptography was done using manual techniques. When it comes to data privacy and security, each and every user or customer wants their personal and confidential data to be secured against malicious or unauthorized access. Major aspects of data or information security are content security, device safety, behavior security and data security. Nowadays, users store or access data from the cloud environment through the internet (Deka and Das, 2014; Namasudra, 2017; Changxiang et al., 2007; Namasudra et al., 2014, 2017a). The rapid growth of the internet has led to an increase in the number of hackers and attackers, and thus, security is becoming a major concern. A strong cryptographic technique can be an appropriate solution, and it must protect users' confidential and sensitive data at any cost. Traditional cryptographic schemes are mainly developed based on theoretical analysis and complex mathematical calculations. However, there are some trapdoors in the traditional schemes, such as RSA (Ron Rivest, Adi Shamir and Leonard Adleman), Data Encryption Standard (DES), etc. So, these cryptographic schemes can be compromised by attacks. Thus, it can be stated that the modern mathematical calculation–based cryptographic schemes are not as reliable as they once were.

Deoxyribonucleic Acid (DNA) cryptography is one of the best solutions to the aforementioned problem. In DNA cryptography, DNA is used as a computational and informational carrier along with molecular technologies (Adleman, 1994). The concept of DNA has received much attention in the field of information security due to its complex structure.

DNA can also be used to store data; one gram of DNA can store about 700 TB of data, which surpasses the storage ability of optical, electrical or magnetic storage media (Cui et al., 2006). That means that a few grams of DNA can store all the world's data. An interdisciplinary knowledge of computer science as well as biology is required to use the concept of DNA in the field of cryptography.

DNA is basically the repository of meaningful information about cells, where a *cell* can be defined as the fundamental unit of life. In 1869, physician Friedrich Miescher first isolated and identified DNA. DNA is a lengthy polymer, and it is built by using nucleotides. There are three main components of nucleotides:

1. A nitrogenous base
2. A five-carbon sugar
3. A phosphate group

There are four main types of nitrogen base: Adenine (A), Guanine (G), Cytosine (C) and Thymine (T). These nitrogen bases are combined with deoxyribose to generate a long sequence. Each string is paired with the complementary string by using the Watson-Crick complement rule for making a double helix. The rule is that G is always paired with C and A is paired with T.

In 1994, DNA was first proposed in the field of computation by Adleman (1994). By using DNA computing, Adleman solved the seven-point "Hamiltonian Path Problem." Since then, much research has been done using DNA computing in many fields. Clelland et al. (1999) proposed a method for hiding messages or data in DNA microdots. A *microdot* can be defined as the process for concealing messages. Barish et al. (2005) have discussed a system that gives output by using DNA. This system is also used for solving many NP complete problems. Rothemund et al. (2004) have demonstrated that the Exclusive OR (EXOR) function, which is one of the important parts of any cryptosystem, can be computed by using DNA. Several schemes have been developed by using DNA cryptography for breaking modern algorithms, such as DES (Boneh et al., 1995), RSA (Beaver, 1994) and Number Theory Research Unit (NTRU) (Pelletier and Weimerskirch, 2002). Nowadays, research is concentrating on using DNA computing for developing new data security schemes, which can be categorized into two types:

1. DNA-based data encryption
2. DNA-based data hiding

In the DNA-based data encryption technique, data are converted into binary form, and then, this binary form of data is converted into a DNA sequence by using a *DNA encoding* method. The complementary pair rule and other approaches are applied to the DNA sequence to improve security. On the other hand, in the DNA-based data hiding technique, data are hidden in a DNA sequence. Based on the aforementioned two categories, a taxonomy of a DNA-based security model is presented in this chapter. In this taxonomy, many existing DNA-based security models are briefly discussed along with their advantages and disadvantages. A taxonomy can be very helpful in developing novel security models using DNA computing.

The rest of the chapter is organized into several sections. Section 3.2 presents the proposed taxonomy for DNA-based security models. The reasons for the proposed taxonomy are given in Section 3.3. Finally, Section 3.4 concludes the chapter.

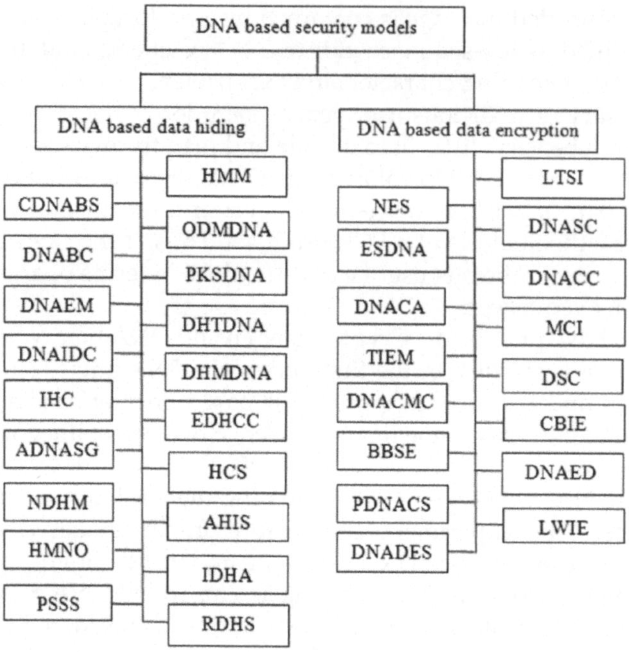

FIGURE 3.1
Taxonomy of DNA-based security models.

3.2 Taxonomy of DNA-Based Data Security Models

Many researchers have proposed many cryptographic schemes by using DNA computing, which can be categorized into two types: DNA-based data hiding schemes and DNA-based data encryption schemes. Figure 3.1 shows a taxonomy of DNA-based security models.

3.2.1 DNA-Based Data Hiding Schemes

Due to the increasing number of hackers and malicious users, data hiding is one of the most popular ways to ensure data protection. In traditional approaches, data were hidden in a cover image, so that they could be protected from attackers. Recently, many researchers have proposed many techniques for data hiding by using DNA computing, which is known as *DNA steganography*. Both steganography and cryptography have a common aim, but use different approaches. The former hides the existence of the data, and the latter changes the content of the data. In steganography, a function is used to identify the existence of data. On the other hand, in cryptography, a secret key is used to convert a plaintext into a ciphertext. Without any prior knowledge of the secret key, no one is able to decrypt the ciphertext to get the plaintext.

Clelland et al. (1999) proposed the first method for data hiding, namely Hiding Messages in Microdots (HMM). Using a microdot, they developed a DNA-based steganographic technique to send secret data or messages. In this scheme, a DNA-encoded message or data is first hidden within the massive complex of human genomic DNA, and then, the encoded message is further concealed by restricting this sample to a microdot. In this scheme, an encoded message is encrypted by Polymerase Chain Reaction (PCR) primer

sequences, and the encoded message is contained by a prototypical secret message. Here, a *primer* can be defined as ten-sequence pair of DNA. Clelland et al. have used a simple substitution cipher for encoding characters in DNA triplets.

Even if an adversary or malicious user gets a microdot, without the respective primer sequences, it will still be very difficult to decode and read the message. For an example, if random 20 base primers are used to enhance the DNA, separate enhancements with more than 10^{20} different pair of primers are required, even if three mismatches per primer are allowed followed by the analysis of a PCR products. Thus, if the same pair of primers is used in many cases, the malicious user faces difficulties in terms of an experimental barrier to get the primer sequences.

Leier et al. (2000) have proposed a Cryptosystem using DNA Binary Strands (CDNABS) that consists of two approaches. In the first approach, DNA binary strands are used for information hiding, and the second approach is based on graphical subtraction, which helps to strengthen the security of the first approach. However, neither of these approaches support fast data accessing. In 2003, Wong et al. (2003) carried out Organic Data Memory using DNA (ODMDNA) experiments for evaluating the feasibility of embedding a message or data in a host. Their experiments contain four primary stages: DNA host identification, information encoding, unique DNA searching and wet laboratory procedures. In the first stage, candidates are identified for carrying an embedded DNA molecule. Wong et al. considered bacteria and plants as message hosts and also considered the stamina of the DNA host. In the information encoding stage, DNA encoding is used to convert the data or message into the DNA bases. In the searching stage, a set of fixed size DNA sequences is identified from the database (www.tigr.org), which does not exist on the host candidates, but satisfies all the restrictions and constraints of genomic. There are four methods in wet laboratory procedures: create complementary oligos, insert DNA, incorporate into genome and extract the message. In the *create complementary oligos* method, a restricted enzyme site is created for encoded DNA fragments, and the fragments of DNA are then cloned in the recombinant plasmid. In the *insert DNA* method, embedded DNA is inserted in the cloning vectors. In the *incorporate into genome* method, the encoded DNA and vector are integrated into the genome for permanent data or message storage. In the *extract the message* method, the message or data part of the DNA strands is extracted from bacteria by using PCR. This scheme does not support fast data accessing.

In a conventional electronic system, the real-time applications of cryptography are limited because of the size of the one-time pad. The *one-time-pad* encryption technique uses a codebook of arbitrary data or files to convert it from plaintext to ciphertext. To solve the above-mentioned problem, Gehani et al. (2000) proposed a scheme called DNA-Based Cryptography (DNABC) for secretly assembling large records of one-time pads in the structure of DNA strands, and then, isolating and cloning them. The use of DNA sequences can mitigate this issue since DNA can store a huge amount of data in a small physical volume. Gehani et al. have assumed that one-time pads are shared by the sender and receiver in advance of the secret message, and they have proposed two DNA one-time-pad encryption methods, namely the substitution method and the EXOR method. In the substitution one-time-pad system, a plaintext binary message and a table are used, which define a random mapping to ciphertext. Here, encryption is executed by substituting each and every plaintext DNA with a corresponding DNA cipher. To implement the EXOR one-time-pad cryptosystem with DNA, there are few methods: encode a plaintext data or message or generate a DNA one-time pad and bitwise EXOR in DNA. Since the XOR operation is its own inverse, the decryption operation of a ciphertext is effected by using the same process

as the encryption operation with the same key. However, if an attacker gets one-time pads, s/he can hack the data.

Tanaka et al. (2005) have proposed a Public Key System using DNA (PKSDNA) and a one-way function. This scheme is mainly developed for solving the key distribution issue, where the messages are hidden in the dummy DNA sequences and restored by using the PCR amplification and sequencing. The following steps are used in this scheme for key distribution:

Step 1: In the first step, the sender selects the public key, and the key is presented at the receiver's end.

Step 2: The sender encodes the data or message by using the DNA synthesizer along with the Public Key (PK) holding the forward primer.

Step 3: In this step, the PK that is holding a reverse primer is tied up.

Step 4: In the fourth step, the dummy sequences are used to conceal the encoded DNA message in the magnetic beads, and then, it is sent to the receiver.

Step 5: The encoded secret key sequences, which are placed between the correct primer sequences, are then enlarged by using the PCR method. Now, the secret key can be exchanged between the sender and receiver by using the sequence analysis.

In this scheme, everyone can make the group of encoded DNA messages, but only the authorized user who has the actual public key can get the message from the group. This scheme is not secure against man-in-the-middle attacks.

In 2006, Hsu and Lee (2006) proposed a DNA-based Encryption Method (DNAEM) consisting of three encryption processes to improve data security: insertion, complementary and substitution. In all these processes, the binary coding rule and complementary pair rule are kept secret by the sender and receiver. The *binary coding rule* transforms the DNA bases into binary codes and vice versa as shown in Table 3.1. In the *complementary pair rule,* the complement of each alphabet (*a*) is denoted as *C*(*a*). The rule: ((A C) (C G) (G T) (T A)) is used as a complementary rule. But, if any hacker or malicious user gets the binary encoding rule and complementary pair rule, there is a chance that they will be able to hack the data or message.

Shiu et al. (2010) have proposed Data Hiding Techniques based on DNA (DHTDNA) by utilizing DNA sequences. In DHTDNA, a DNA reference sequence is selected, and confidential data or a message is incorporated into it. Then, it is sent to the receiver. Data may face a security issue in DHTDNA, if any attacker gets the DNA reference key. Kencla and Loebl (2010) discussed the data hiding processes and proposed a DNA Inspired Data Concealing (DNAIDC) method. This scheme is easy to implement.

TABLE 3.1

An Example of Binary Coding

DNA Bases	Binary Code
A	00
C	01
G	10
T	11

In 2011, Abbasy et al. proposed a Data Hiding Method using DNA (DHMDNA) for increasing the complexity (Abbasy et al., 2011). Their scheme is based on the complementary pair rule and binary coding. There are two main phases in this scheme:

1. Embedding secret message
2. Extracting original message

In the embedding phase, a DNA reference sequence is shared between the sender and the receiver from the National Center for Biotechnology Information (NCBI) or European Bioinformatics Institute (EBI) database. Figure 3.2 shows a block diagram of the embedding secret message phase. The extracting phase is the reverse process of the embedding phase for extracting the original message.

Torkaman et al. (2011) proposed an Improving Hybrid Cryptosystem (IHC), where both symmetric and asymmetric cryptography are used for exchanging secret messages. Here, a DNA-based steganographic algorithm is used to hide a secret message. There are two main rules in the DNA steganography process of IHC:

1. DNA coding technology rule
2. DNA complementary rule

The DNA coding technology rule is kept secret between the sender and the receiver, and using this rule, binary data is converted to DNA string. The DNA complementary rule is also kept secret between the sender and the receiver, and for each DNA sequence, there are six complementary rules, as shown in Table 3.2. In both DHMDNA and IHC, if any attacker gets the DNA reference key, s/he can get the secret data.

Abbasy and Shanmugam (2011) proposed a novel scheme called Enabling Data Hiding for Cloud Computing (EDHCC). The main aim of EDHCC is to increase the complexity of the algorithm, where a DNA reference sequence has been selected and confidential data or a message is hidden in it. To maintain the trade-off between cost and security, Sureshraj and Bhaskaran (2012) proposed a novel Automatic DNA Sequence Generation (ADNASG) algorithm for the multi-cloud storage system for securing the data and ensuring its availability. This algorithm is based on the complementary pair rules and binary coding, and

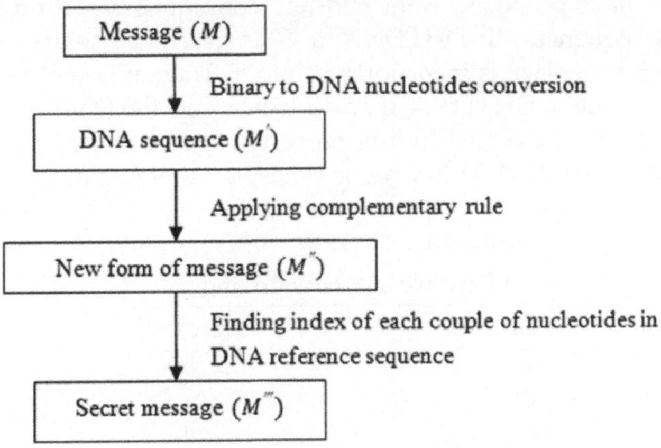

FIGURE 3.2
Embedding phase of DHMDNA.

TABLE 3.2
Six Main Complementary Rules

AT	TC	CG	GA
AT	TG	GC	CA
AC	CT	TG	GA
AC	CG	GT	TA
AG	GT	TC	CA
AG	GC	CT	TA

a DNA reference sequence is selected to hide secret data M. Then, the encrypted data are uploaded on the cloud server. Cloud servers offer reduced IT costs, business continuity, unlimited storage, etc. (Deka et al., 2013; Deka and Borah, 2012; Namasudra and Roy, 2015, 2016, 2017a–c, 2018; Namasudra et al., 2017b; Sarkar et al., 2015; Namasudra, 2018; Majumder et al., 2014). To provide a high level of privacy as well as to ensure the data's availability, Sureshraj and Bhaskaran divided the users' data into data pieces, which are then distributed among the available service providers based on the users' budgets. When the clients or users decide to use the data, it is retrieved from the DNA reference sequence. Dividing and distributing users' data can easily ensure the data's or files' availability on the cloud server instead of misusing the user's confidential data. Figure 3.3 shows the flowchart of the system model of ADNASG. EDHCC and ADNASG do not support fast data accessing, and data security may be at risk, if an attacker gets fundamental information like the DNA sequence, complementary rules, etc.

Khalifa and Atito (2012) proposed a High Capacity Steganographic (HCS) scheme using the basic properties of DNA. There are two main steps in this scheme. In the first step, a DNA-based playfair cipher is used to encrypt the message. In the second step, a substitution process is applied to conceal the encrypted message in some DNA references. Khalifa and Atito changed the one-by-one base pair complementary rule to a two-by-two complementary rule, where sender and receiver share a DNA-based reference key. Nowadays, encrypted messages or data are embedded into multimedia files, such as audio, image, video, etc. (Park et al., 2009). In 2013, a Novel Data Hiding Method (NDHM) was proposed, wherein the secret message is embedded into a Word document (Liu et al., 2013). Here, at

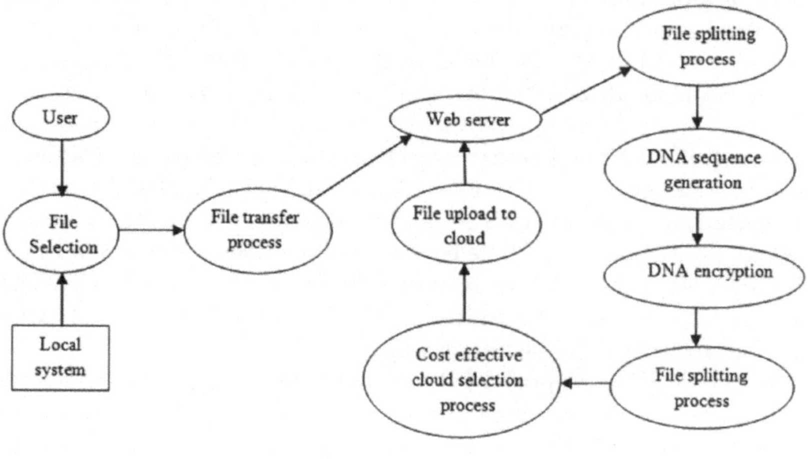

FIGURE 3.3
Flowchart of the system model of ADNASG.

first, the plaintext of the data is embedded into a DNA sequence. Then, the DNA sequence is again encrypted by another DNA sequence of the same length, which is generated by using Chebyshev maps, and the outcome is appended to a primer of the DNA sequence. Since there are many steps in NDHM, it may take a long time to encrypt a single data.

In 2015, a few schemes were proposed to hide data by using DNA computing (Roy et al., 2015; Zhang and Gao, 2015; Gupta and Singh, 2015). Roy et al. (2015) proposed an Approach for Hybrid Image Steganography (AHIS), where three layers of steganographic data encryption processes are presented. In the first layer, secret data are encoded in a DNA sequence. In the second layer, the authentication values are used, and the data are encrypted by using a stego-key inside the DNA sequence. In the third layer, data are encoded inside an image file along with the authentication key. This encrypted image is called a *stego-image*. In the decryption process, the encrypted data are extracted from the stego-image by using the proper authentication value, and then, the extracted bit is decrypted using the stego-key. Then, the original data are retrieved from the sequence.

Zhang and Gao (2015) proposed a novel method for data hiding called *Hiding data by Module-N Operation* (HMNO) based on DNA encoding and substitution. Here, to achieve the substitution, the secure information and the cover media are encoded. A random grouping algorithm is proposed in HMNO for generating the pseudo-genetic codon table that regulates the substitution. In DNA, *codons* can be defined as a group of three neighboring nucleotides that corresponds to any amino acid or stop signal. In this scheme, the cover media is encoded into a sequence like DNA (pseudo DNA sequence). Then, the sequence is rearranged into a sequence like a codon (pseudo-codon sequence). After this process, a table of pseudo-genetic codons is constructed for mapping the pseudo-codon sequence with different groups. Each codon of the pseudo-codon sequence is allied to one group of the pseudo-codon table, and the total number of codons in one group is random as per the randomly constructed table of pseudo-genetic codons. Therefore, an arbitrary N-nary system is attained with the pseudo-genetic codon sequence. Secure embedded data is transformed into N-nary form in this scheme by using the module-N operation as per the construction of arbitrary N-nary system. Finally, the pseudo codons in the pseudo-codon sequences are replaced with the codons of the same group as per the N-nary information.

Gupta and Singh (2015) proposed an Improved Data Hiding Approach (IDHA) using DNA computing. There are two main processes in IDHA: a substitution process, and the central dogma of molecular biology. In the substitution process, a DNA sequence is randomly taken from a publicly available DNA database. In the second process, the DNA sequence is transformed into a ribonucleic acid sequence, and then, it is transformed into a protein sequence. Here, three DNA bases are used to encrypt data, so data security may be at risk.

Based on the secret sharing scheme (Wang et al, 2007), Tuncer and Avci (2016) proposed a novel Probabilistic Secret Sharing Scheme (PSSS) by using a DNA-EXOR operator. In PSSS, a truth table is used based on DNA-EXOR, and the data are divided into three secret shares by using a DNA-EXOR operator, where each part of the data is embedded into red, green and blue channels of the cover image, respectively. Table 3.3 shows DNA-EXOR rules. But, PSSS may not be secured against man-in-the-middle attacks, since if any attacker gets the truth table, s/he can retrieve the secret message.

In 2017, Wang et al. (2017) proposed a Reversible Data Hiding Scheme (RDHS) by using the DNA-EXOR operation rule, where information is embedded, and after the extraction process, is restored exactly as the original host signal. RDHS is inspired by the algorithm of Ni et al. (2006) for histogram modification. There are two main algorithms in RDHS, namely embedding watermarking and extracting watermarking.

TABLE 3.3

DNA-EXOR Rules

EXOR	A	G	C	T
A	A	G	C	T
G	G	A	T	C
C	C	T	A	G
T	T	C	G	A

In the embedding process, using the rule of Ni et al., the watermarking is encoded into the DNA sequence, and then, embedding watermarking integrates the max base along with the watermarking base using a DNA-based EXOR operation rule. The max bases of the cover image are swapped with the results of the EXOR operation.

The extracting watermarking process is the reverse process of the embedding process. First, the max base and the location are obtained. In the second step, watermarking is extracted by using the DNA-based EXOR operation rule. Then, the last bases of the watermarked image are replaced or swapped with the results of the second step, and the watermarking and the image after extracting the watermarking are decoded. Finally, the watermarking and image are output. However, if any attacker gets the binary encoding rule and DNA-EXOR operation rule in RDHS, s/he can get the original content of the data.

3.2.2 DNA-Based Data Encryption Schemes

DNA-based encryption techniques are among the most popular data encryption techniques. Many researchers have proposed many data security schemes by using DNA computing. In this subsection, many DNA-based data encryption schemes are discussed.

In 2001, Bancroft et al. (2001) proposed two techniques for Long Term Storage of Information (LTSI): DNA sequence analysis and PCR. Their scheme uses two classes of DNA: information DNA (iDNA), which contains information; and a Poly Primer Key (PPK), which is responsible for retrieving the information from iDNA. Each and every iDNA keeps a few elements, namely regular adjoining Forward (F) and reverse (R) PCR enlargement primers, a primer, a common spacer and a unique information segment. To retrieve the information, PPK is first enlarged by using F and R primers. PCR enlargement yields sufficient amounts of PPK. Then, sequence analysis of PPK reveals the sequences of F and R primers. This analysis leads to the information being retrieved. Amin et al. (2006) proposed a Novel Encryption Scheme (NES) by using DNA computing. Here, binary forms of secret data or messages are converted into DNA sequences. Then, an efficient searching algorithm is used to locate the positions of the DNA bases for each plaintext character. However, neither LTSI and NES schemes support strong data security.

In 2007, MingXin et al. (2007) proposed a symmetric cryptosystem called *DNA Symmetric-key Cryptosystem* (DNASC). Here, both the encryption and decryption keys are generated by using DNA computing. In DNASC, a huge number of DNA probes are hybridized and identified at the same time, and thus, the decryption processes are executed in a parallel way. But, this scheme takes a long time to encrypt and decrypt any data. Guangzhao et al. (2008) proposed an Encryption Scheme using DNA (ESDNA) computing, which involves synthesizing method, DNA digital coding, PCR amplification and traditional cryptography methods. The coding mode and primers are used as the key in their scheme. Here, DNA digital coding and traditional encryption techniques are used to pre-process the plaintext. There are five main phases in this scheme: key generation, data pre-processing,

encryption, decryption and data post-processing. In the key-generation phase, the sender and the receiver of the message exchange and design PCR pair primers. Here, the encryption and decryption keys are PCR pair primers. In the data pre-processing phase, the plaintext of the data is converted into hexadecimal code, and then, the hexadecimal code is converted into binary form. In encryption phase, the binary data is encrypted by using the public key of the receiver, and then, it is converted into the sequence using the technology of DNA digital coding. Finally, a DNA sequence holding an encoded message or data is flanked by using the reverse and forward PCR primers. The sender sends the message to the receiver after mixing it with some dummy sequences. In the decryption phase, the receiver uses the primer pairs to get the secret DNA message sequence, and then, uses DNA digital coding method and secret key to get the data in the binary plaintext. In the data post-processing phase, the receiver retrieves the plaintext from the binary plaintext using a post-treatment technique. However, this scheme is slow for data accessing.

Wang and Zhang (2009) proposed a DNA Computing Based Cryptography (DNACC) model by using RSA algorithm. DNACC consists of five phases: setup plaintext, matching with nucleotides, converting nucleotides to numbers, encrypted message and recovery plaintext. In the first phase, the sender of a message sets up the plaintext for sending the message. In the matching with nucleotides phase, the plaintext is matched with the nucleotides by using Table 3.4. In the third phase, nucleotides are converted to numbers. In the encrypted message phase, the message is encrypted by using an asymmetric key. In the last phase, the receiver recovers the message. In this scheme, if any attacker gets the large prime numbers, then, s/he can easily get the original content of the data. Tornea and Borda (2009) proposed DNA Cryptographic Algorithms (DNACA) for improving data security. In their first algorithm, public key encryption has been used with the RSA algorithm, and then, encrypted data have been transformed into the DNA sequence. In the second algorithm, the EXOR operation has been used to transform the binary data into DNA structures. But, this scheme is vulnerable to attacks. Shoshani et al. (2012) proposed a cryptosystem by using DNA computing, namely Molecular Cryptosystem for Image (MCI). They have mainly concentrated on the retrieval process of secure images by using DNA molecules, but MCI increases the system overhead.

In 2013, Babaei (2013) proposed an efficient method for encryption, namely the Text and Image Encryption Method (TIEM). TIEM is mainly based on DNA computing and chaos theory, and a one-time-pad algorithm has been used in TIEM to enhance performance. A symmetric-key cryptography method has been proposed by Kar et al. (2013) using a DNA sequence, namely Data Security and Cryptography (DSC). Here, instead of sending the actual key directly, the sender and the receiver share a session key, which contains information from the encryption key. Both TIEM and DSC schemes are not efficient, when users need to access the data in a shorter time.

Ranalkar and Phulpagar (2014) proposed a novel DNA Cryptography in Multi-Cloud (DNACMC) scheme for the multi-cloud environment, where different cloud service providers meet together. DNACMC consists of two phases: data embedding and data extracting.

TABLE 3.4

Rules for Matching the Alphabet with Nucleotides

Alphabet	Nucleotide	Alphabet	Nucleotide
A	TT	.	.
B	AC	.	.
C	AG	Z	GT

In data embedding phase, first a data or message M is converted into binary form by the sender. Then, after applying binary coding, a DNA sequence M' is created, and then, M' is converted to M'' after applying the base pair rule. In the next step, an index of nucleotides is found in the DNA sequence and made into a final ciphertext as M'''. The sender then uploads the ciphertext M''' on the cloud server. The decryption phase is the reverse of the encryption phase. Here, the receiver first downloads M''' from the cloud server, and then, after searching an index of the nucleotides from the DNA sequence, M''' is converted into M''. Then, the receiver uses the base pair rule and binary coding to retrieve the original message M. This scheme is not secured against man-in-the-middle attacks. Enayatifar et al. (2014) proposed a Chaos-Based Image Encryption (CBIE) scheme based on genetic algorithm, logistic map and DNA masking. The main goal of CBIE is to find the best DNA mask for image encryption. In the first phase of CBIE, many DNA masks are created by using the DNA sequences and logistic map. In the next phase, a genetic algorithm is used to find the best DNA mask. Table 3.5 shows the encoding and decoding map rules used in CBIE for DNA sequences. Jain and Bhatnagar (2014) proposed the Bit Based Symmetric Encryption (BBSE) scheme by using DNA sequence and complementary pair rules for improving data security. They have mainly tried to increase the complexity of the key and complementary rule. In BBSE, instead of passing the actual information through the intermediate medium, only the complement of the key is passed. However, this scheme does not provide strong security for users' confidential data.

In 2015, a DNA-based Encryption and Decryption (DNAED) scheme was proposed (Rahman et al., 2015). There are two main phases in this scheme: (1) a DNA computing-based encoding algorithm, and (2) a DNA computing–based encryption and decryption algorithm. In the first phase, a plaintext is converted into DNA sequences by using a DNA encoding rule. Then, in the next phase, the sender encrypts the DNA sequences and sends them to the receiver for decryption. The execution process of this scheme is slow.

Kalyani and Gulati (2016) proposed a Pseudo DNA Cryptography Scheme (PDNACS), which provides three levels of security by using the DNA complementary rule, arithmetic operation (i.e., EXOR operation) and one-time pad. There are many phases in this scheme, namely binary conversion, one-time pad, EXOR operation, DNA conversion and DNA encryption. So, data encryption takes a long time. Mondal and Mandal (2016) proposed a Light Weight Image Encryption (LWIE) scheme based on the logistic map and DNA sequencing. Here, the plain image is permuted by using a sequence of pseudo-random numbers and is then encrypted by using the DNA bases. The substitution phase of LWIE makes the scheme lightweight, and the permutation process is carried out on the plain image for better performance and quality. LWIE has been proposed only for gray-label

TABLE 3.5

Encoding and Decoding Map Rules of CBIE

Rule	A	T	C	G
Rule 1	00	11	10	01
Rule 2	00	11	01	10
Rule 3	11	00	10	01
Rule 4	11	00	01	10
Rule 5	10	01	00	11
Rule 6	01	10	00	11
Rule 7	10	01	11	00
Rule 8	01	10	11	00

images. In 2017, a DNA Data Encryption Standard (DNADES) scheme was proposed by Ahmed and Henawy (2017). DNADES is based on DNA, nucleic acid and ribonucleic acid for increasing the available number of permutations of traditional DES and for increasing the key space. This scheme is more robust than traditional DES. However, the data accessing time is high in this scheme.

3.3 Reasons for Taxonomy

All DNA-based security models have their own advantages and disadvantages. A security model that performs efficiently in one working environment may not be suitable in another environment. Therefore, an efficient security model must be selected at the time of communicating users' personal or confidential data. As discussed earlier, DNA-based security models are very popular nowadays because of the method's complex structure. Classification or taxonomy of DNA-based security models helps the system developers to understand the work process of a particular model along with its advantages and disadvantages. Thus, it can lead to the accurate selection of a DNA-based security model for a particular work environment.

If there are many attackers and malicious users in an environment and the system administrator wants to encrypt the users' confidential data, DNA-based encryption models can be useful. In the taxonomy presented here, many DNA-based encryption models have been briefly discussed. DNA encoding, primers, codons and complementary pair rules are used in DNA-based encryption models to improve data security. Some models do not provide much security, whereas, some models take a long time to encrypt data or messages because of the complex operation involved. So, if a system administrator does not want to consider time, the taxonomy can be used to select a better DNA-based security model for users' confidential or sensitive data.

On the other hand, in DNA-based data hiding models, the plaintext of the data is converted into the DNA sequence by using a DNA computing technique, and it is processed to convert it into some other complex structure. Finally, the existence of the DNA sequence is hidden in some other fake DNA sequences. If a system administrator wants to hide the existence of the data, then, by using the taxonomy, a suitable model can be selected by analyzing its advantages and disadvantages.

Thus, it is clear that the taxonomy of DNA-based security models can be very useful for selecting an appropriate security model in a given environment.

3.4 Conclusions

Nowadays, DNA computing is one of the popular computing fields that is used for improving data security. DNA computing is popular because of the complex structure of DNA and the storage capacity it offers. There are several existing security models based on DNA computing. In this chapter, a taxonomy of DNA-based security models has been presented based on DNA-based encryption techniques and DNA-based data hiding techniques. All the existing DNA-based security schemes are discussed under the taxonomy along with

their advantages and disadvantages. The taxonomy of DNA-based security models can be very helpful for developing a novel DNA-based security model in the future.

Key Terms and Definitions

Steganography: Steganography involves hiding the existence of data. In DNA steganography, data are hidden by using a DNA sequence. Here, data are first converted into binary form, and then, by using some base pairing rule, the binary data is converted into a DNA sequence.

Binary coding: In DNA computing, the binary coding rule transforms the DNA bases, namely A, C, G and T, into binary codes and vice versa. For example, binary coding rule ((A 00) (C 01) (G 10) (T 11)) or ((A 01) (C 00) (G 11) (T 110)) or some other rule can be used to transform the DNA bases into binary codes.

Complementary pair rule: In molecular biology, complementarity is achieved by the interactions between the bases, namely A, T, G and C. These bases bond together in nucleic acid through hydrogen bonding. The base pair G–C forms three hydrogen bonds and the A–T pair forms two hydrogen bonds. A complementary DNA strand can be constructed based on complementarity. In the complementary pair rule, the complementary of each alphabet x is denoted as $C(x)$. The rule: ((A C) (C G) (G T) (T A)) can be used as a complementary pair rule.

References

Abbasy, M. R. and Shanmugam, B. (2011). Enabling data hiding for resource sharing in cloud computing environments based on DNA sequences. In *Proceedings of the 2011 IEEE World Congress on Services*, IEEE, Washington, DC, pp. 385–390.

Abbasy, M. R., Manaf, A. A., and Shahidan, M. A. (2011). Data hiding method based on DNA basic characteristics. In Ariwa, E. and Qawasmeh, E. E. (Eds.), *Digital Enterprise and Information Systems*, (pp. 53–62). Springer: Berlin Heidelberg.

Adleman, L. (1994). Molecular computation of solutions to combinatorial problems. *Science*, 266 (5187), 1021–1024.

Ahmed, K. and Henawy, I. E. (2017). Increasing robustness of data encryption standard by integrating DNA cryptography. *International Journal of Computers and Applications*, 39(2), 91–105.

Amin, S. T., Saeb, M., and Gindi, S. E. (2006). A DNA-based implementation of YAEA encryption algorithm. In *Proceedings of the 2nd IASTED International Conference on Computational Intelligence*, San Francisco, CA.

Babaei, M (2013). A novel text and image encryption method based on chaos theory and DNA computing. *Natural Computing*, 12(1), 101–107.

Bancroft, C., Bowler, T., Bloom, B., and Clelland, C. T. (2001). Long-term storage of information in DNA. *Science*, 293 (5536), 1763–1765.

Barish, R. D., Rothemund, P. W., and Winfree, E. (2005). Two computational primitives for algorithmic self-assembly: Copying and counting. *Nano Letters*, 5(12): 2586–2592.

Beaver, D. (1994). Factoring: The DNA solution. In *Proceedings of the 4th International Conference on the Theory and Applications of Cryptology, Wollongong*, Australia, pp. 419–423.

Boneh, D., Dunworth, C, and Lipton, R. (1995). Breaking DES using a molecular computer. In *Proceedings of the DIMACS Workshop on DNA Computing*, pp. 37–65.

Changxiang, S., Zhang, H., Feng, D., Cao Z., and Huang, J. (2007). Survey of information security. *Science in China Series F: Information Sciences*, 50(3), 273–298.

Clelland, C. T., Risca, V., and Bancroft, C. (1999). Hiding messages in DNA microdots. *Scientific Correspondence, Nature*, 399, 533–534.

Cui, G., Liu, Y., and Zhang, X. (2006). New direction of data storage: DNA molecular storage technology. *Computer Engineering and Application*, 42(26): 29–32.

Deka, G. C. and Borah, M. D. (2012). Cost benefit analysis of cloud computing in education. In *Proceedings of the International Conference on Computing, Communication and Applications*. IEEE, Dindigul, Tamilnadu, India, pp. 1–6.

Deka, G. C. and Das, P. K. (2014). An overview on the virtualization technology. *Handbook of Research on Cloud Infrastructures for Big Data Analytics*, IGI Global. DOI: 10.4018/978-1-4666-5864-6.ch012.

Deka, G. C., Kathing, M., and Kumar, D. P. (2013). Library automation in cloud. In *Proceedings of the International Conference on Computational Intelligence and Communication Networks*. IEEE, Mathura.

Enayatifar, R., Abdullah, A. H., and Isnin, I. F. (2014). Chaos-based image encryption using a hybrid genetic algorithm and a DNA sequence. *Optics and Lasers in Engineering*, 56, 83–93.

Gehani, A., LaBean, T., and Reif, J. (2000). DNA-based cryptography. In Jonoska, N., Paun, G., and Rozenberg, G. (Eds.), *Aspects of Molecular Computing*. (pp. 167–188). Springer: Berlin Heidelberg.

Guangzhao, C., Qin, L., Wang, Y., and Zhang, X. (2008). An encryption scheme using DNA technology. In *Proceedings of the 3rd IEEE International Conference on Bio-Inspired Computing: Theories and Applications, BICTA*, IEEE, Adelaide, SA, Australia, pp. 37–42.

Gupta, R. and Singh, R. K. (2015). An improved substitution method for data encryption using DNA sequence and CDMB. In *Proceedings of the International Symposium on Security in Computing and Communication*, Kochi, India, pp. 197–206.

Hsu, H. Z. and Lee, R. C. T. (2006). DNA based encryption methods. In *Proceedings of the 23rd Work-Shop on Combinatorial Mathematics and Computation Theory*, Nantou Hsies, Taiwan, pp. 145–150.

Jain, S. and Bhatnagar, V. (2014). Bit based symmetric encryption method using DNA sequence. In *Proceedings of the 5th International Conference on Confluence*, IEEE, Noida, India, pp. 495–498.

Kalyani, S. and Gulati, N. (2016). Pseudo DNA cryptography technique using OTP key for secure data transfer. *International Journal of Engineering Science and Computing*, 6(5), 5657–5663.

Kar, N., Majumder, A., Saha, A., and Deb, S. (2013). Data security and cryptography based on DNA sequencing. *International Journal of Information Technology and Computer Science*, 10(3), 24–32.

Kencla, L. and Loebl, M. (2010). DNA-inspired information concealing: A survey. *Computer Science Review*, 4(4), 251–262.

Khalifa, A. and Atito, A. (2012). High-capacity DNA-based steganography. In *Proceedings of the 8th International Conference on Informatics and Systems*, IEEE, Cairo, Egypt, pp. 76–80.

Leier, A., Richter, C., Banzhaf, W., and Rauhe, H. (2000). Cryptography with DNA binary strands. *BioSystems*, 57(1), 13–22.

Lin, H. S. (1998). Cryptography and public policy. *Journal of Government Information*, 25(2), 135–148.

Liu, H. J., Lin, D., and Kadir, A. R. (2013). A novel data hiding method based on deoxyribonucleic acid coding. *Computers and Electrical Engineering*, 39, 1164–1173.

Majumder, A., Namasudra, S. and Nath, S. (2014). Taxonomy and classification of access control models for cloud environments. In Mahmood, Z. (Ed.), *Continued Rise of the Cloud*, (pp. 23–53). Springer: London.

MingXin, L., XueJia, L., GuoZhen, X, and Lei, Q. (2007). Symmetric-key cryptosystem with DNA technology. *Science in China Series F: Information Sciences*, 50(3), 324–333.

Mondal, B. and Mandal, T. (2016). A light weight secure image encryption scheme based on chaos and DNA computing. *Journal of King Saud University – Computer and Information Sciences*, 29(4), 499–504.

Namasudra, S. (2017). An improved attribute-based encryption technique towards the data security in cloud computing. *Concurrency and Computation: Practice and* Exercise. DOI:10.1002/cpe.4364.

Namasudra, S. (2018). Cloud computing: A new era. *Journal of Fundamental and Applied Sciences*, 10(2), 113–135.

Namasudra, S. and Roy, P. (2015). Size based access control model in cloud computing. In *Proceedings of the International Conference on Electrical, Electronics, Signals, Communication and Optimization*, IEEE, Visakhapatnam, India, pp. 1–4.

Namasudra, S. and Roy, P. (2016). Secure and efficient data access control in cloud computing environment: A survey. *Multiagent and Grid Systems – An International Journal*, 12(2) 69–90.

Namasudra, S. and Roy, P. (2017a). Time saving protocol for data accessing in cloud computing. *IET Communications*, 11(10), 1558–1565.

Namasudra, S. and Roy, P. (2017b). A new secure authentication scheme for cloud computing environment. *Concurrency and Computation: Practice and Exercise*, 29(20). DOI:10.1002/cpe.3864.

Namasudra, S. and Roy, P. (2017c). A new table based protocol for data accessing in cloud computing. *Journal of Information Science and Engineering*, 33(3), 585–609.

Namasudra, S. and Roy, P. (2018). PpBAC: Popularity based access control model for cloud computing. *Journal of Organizational and End User Computing*, 30(4), 14–31.

Namasudra, S., Nath, S., and Majumder, A. (2014). Profile based access control model in cloud computing environment. In *Proceedings of the International Conference on Green Computing, Communication and Electrical Engineering*, IEEE, Coimbatore, India, pp. 1–5.

Namasudra, S., Roy, P., and Balamurugan, B. (2017a). Cloud computing: Fundamentals and research issues. In *Proceedings of the 2nd International Conference on Recent Trends and Challenges in Computational Models*, IEEE, Tindivanam, India.

Namasudra, S., Roy, P., Balamurugan, B., and Vijayakumar, P. (2017b). Data accessing based on the popularity value for cloud computing. In *Proceedings of the International Conference on Innovations in Information, Embedded and Communications Systems (ICIIECS)*, IEEE, Coimbatore, India.

Ni, Z., Shi, Y. Q., Ansari, N., and Su, W. (2006). Reversible data hiding. *IEEE Transactions on Circuits and Systems for Video Technology*, 16(3), 354–362.

Park, B., Park, J., and Lee, S. (2009). Data concealment and detection in Microsoft Office 2007 files. *Digital Investigation*, 5(3–4), 104–114.

Pelletier, O. and Weimerskirch, A. (2002). Algorithmic self-assembly of DNA tiles and its application to cryptanalysis. In *Proceedings of the Genetic and Evolutionary Computation Conference*, New York, NY pp. 139–146.

Rahman, N. H. U., Balamurugan, C., and Mariappan, R. (2015). A novel DNA computing based encryption and decryption algorithm. *Procedia Computer Science*, 46, 463–475.

Ranalkar, R. H. and Phulpagar, B. D. (2014). DNA based cryptography in multi-cloud: Security strategy and analysis. *International Journal of Emerging Trends and Technology in Computer Science*, 3(2), 189–192.

Rothemund, P. W. K., Papadakis, N, and Winfree, E. (2004). Algorithmic self-assembly of DNA Sierpinski triangles. *PLoS Biology*, 2(12), 2041–2053.

Roy, S., Sadhukhan, S., Sadhu, S., and Bandyopadhyay, S. K. (2015). A novel approach towards development of hybrid image steganography using DNA sequences. *Indian Journal of Science and Technology*, 8(22), 1–7.

Sarkar, S., Saha, K., Namasudra, S., and Roy, P. (2015). An efficient and time saving web service based android application. *SSRG International Journal of Computer Science and Engineering (SSRG-IJCSE)*, 2(8), 18–21.

Shiu, H. J., Ng, K. L., Fang, J. F., Lee, R. C. T., and Huang, C. H. (2010). Data hiding methods based upon DNA sequences. *Information Sciences*, 180(1), 2196–2208.

Shoshani, S., Piran, R., Arava, Y., and Keinan, E. (2012). A molecular cryptosystem for images by DNA computing. *Angewandte Chemie – International Edition*, 51(12), 2883–2887.

Sureshraj, D. and Bhaskaran, V. M. (2012). Automatic DNA sequence generation for secured cost-effective multi-cloud storage. In *Proceedings of the International Conference on Software Engineering and Mobile Application Modelling and Development*, pp. 1–6.

Tanaka, K., Okamoto, A., and Saito I., (2005), Public-key system using DNA as a one-way function for key distribution, *Biosystems*, 81(1), 25–29.

Torkaman, M. R. N., Nikfard, P., Kazazi, N. S., Abbasy, M. R., and Tabatabaiee, S. F. (2011). Improving hybrid cryptosystems with DNA steganography. *Communications in Computer and Information Science*, 194, 42–52.

Tornea, O. and Borda, M. E. (2009). DNA cryptographic algorithms. In *Proceedings of the International Conference on Advancements of Medicine and Health Care through Technology*, Springer: Berlin, Heidelberg, pp. 223–226.

Tuncer, U. and Avci, E. (2016). A reversible data hiding algorithm based on probabilistic DNA-XOR secret sharing scheme for color images. *Displays*, 41, 1–8.

Wang, B., Xie, Y., Zhou, S., Zhou, C., and Zheng, X. (2017). Reversible data hiding based on DNA computing. *Computational Intelligence and Neuroscience*, 2017, 1–9. DOI:https://doi.org/10.1155/2017/7276084.

Wang, D., Li, X., and Yi, F. (2007). Probabilistic (n,n) visual secret sharing scheme for grayscale images. In *Proceedings of the International Conference on Information Security and Cryptology*, Springer, pp. 192–200.

Wang, X. and Zhang, Q. (2009). DNA computing-based cryptography. In *Proceedings of the 4th IEEE International Conference on Bio-Inspired Computing*, pp. 1–3.

Willett, M. (1982). Cryptography old and new. *Computers and Security*, 1(2), 177–186.

Wong, P. C., Wong, K. K., and Foote, H. (2003). Organic data memory using the DNA approach. *ACM*, 46(1), 95–98.

Zhang, S. and Gao, T. (2015). A novel data hiding scheme based on DNA coding and module-N operation. *International Journal of Multimedia and Ubiquitous Engineering*, 10(4), 337–344.

References for Advanced/Further Reading

Chang, W. L. (2012). Fast parallel DNA-based algorithms for molecular computations: Quadratic congruence and factoring integers. *IEEE Transactions on NanoBioscience*, 11(1), 62–69.

Danziger, M., and Henriques, M. A. A. (2012). Computational intelligence applied on cryptography: A brief review. *IEEE Latin America Transactions*, 10(3), 1798–1810.

Namasudra, S., Roy, P., Vijayakumar, P., Audithan, S., and Balamurugan, B. (2017c). Time efficient secure DNA based access control model for cloud computing environment. *Future Generation Computer Systems*, 73, 90–105.

4

Novel Data Encryption Scheme Using DNA Computing

Kintaali Abhimanyu Kumar Patro and Bibhudendra Acharya

CONTENTS

4.1 Introduction

With advancements in the field of information services and network technologies, huge amounts of information are stored and transmitted over the internet. As a result, privacy and security of information have become serious issues in the present scenario. In most processes, images are widely used and transmitted over the internet. Hence, the privacy and the security of image data represent a challenging task, aiming to protect against unauthorised access by attackers. The most suitable technique for securing images is an *encryption methodology*. The encryption methodology requires a suitable image encryption algorithm along with secret keys to perform an encryption operation.

A suitable image encryption algorithm should follow the following points:

- The keys should be secure and very sensitive to the algorithm.
- The image encryption techniques should have the ability to access the pixels from the original images to perform the image encryption and decryption operation.
- The image encryption techniques should have strong confusion and diffusion architecture so that the resultant ciphertext cannot be easily attacked.
- The image encryption and decryption operation should be faster so that the encryption and decryption outputs are available quickly.
- The image encryption and decryption should be lossless in nature.

Various traditional image encryption techniques, such as DES (Data Encryption Standard) [1, 2], 3DES, AES (Advanced Encryption Standard) [3], RSA, etc., are used to encrypt and decrypt images. In 2000, Dang and Chau [4] developed an image compression and encryption scheme to secure images over the internet. In this scheme, the authors used DWT (Discrete Wavelet Transform) to perform the image compression operation and DES to perform the image encryption operation. This technique improves the image transmission rate and also enhances transmission security. However, the DES algorithm is very complicated and involves large computations. Hence it is a slower process and is not suitable for processing large amounts of image data [5]. To overcome these problems, an AES algorithm is used. The hardware architecture of the AES algorithm can have high throughput (fast processing of image data) [6, 7]. Several image encryption techniques have been developed by using the AES algorithm. Zeghid et al. [5] proposed an image encryption technique that involved modifying the AES algorithm. Kamali et al. [8] proposed an image encryption method by using a modified version of AES.

Nevertheless, these traditional algorithms are not very suitable for encrypting images because of some intrinsic features of images, such as their large size, bulky data capacity, high correlation among adjacent pixels, high redundancy, etc. [9–12]. So, in practical terms, these techniques have a low level of efficiency in the image encryption field [9].

To meet the challenges of low-level efficiency in image encryption, chaotic map techniques were later developed for image encryption, providing higher efficiency and strong security. Chaotic systems have some prominent features such as non-periodicity,

ergodicity, determinacy, system parameter and initial condition sensitivity, etc. Encryption algorithms require these features to make them more secure and stronger.

In 1989, Matthews [13] proposed the first chaos-based image encryption technique. Since then, researchers have used chaos-based image encryption techniques to design secure cryptosystems. Basically, two groups of chaotic maps are used in image encryption. The first is chaotic maps that are one-dimensional, and the second is chaotic maps that are high-dimensional [14]. The chaotic maps based on one dimension are suitable for image encryption due to their simpler structure, high efficiency, and limited requirement of hardware resources. But due to their complex structure, high-dimensional chaotic maps are less suitable for use in image encryption, because the complex structure increases the hardware resource requirements, which, in turn, increases the cost of the product [15].

Many chaos-based image encryption techniques have been proposed by various researchers. Most of the techniques follow permutation and diffusion architecture. Fridrich [16] first suggested a permutation and diffusion architecture based on an image encryption model. Subsequently, many researchers followed this permutation and diffusion principle to encrypt digital images. In 2011, Wang et al. [17] proposed a combined permutation and diffusion operation based on a fast image encryption algorithm. In this algorithm, both the permutation and diffusion operations are used simultaneously to get a better encryption effect. In 2014, Zhang et al. [18] proposed a single round permutation and diffusion operation based on chaotic image encryption. The main point of using this image encryption technique is to overcome the limitations of using multi-round permutation and diffusion operations in image encryption. In 2017, Li et al. [19] proposed a hyper-chaos-based image encryption technique. In this technique, both the pixel-level and the bit-level permutation operations are used to improve security in the cryptosystem.

Nowadays, to obtain better security, DNA cryptography is used in image encryption. In DNA cryptography, DNA is used as an information carrier. DNA cryptography methods have certain unique distinctions and advantages over other cryptography methods such as traditional cryptography and quantum cryptography. Some of the advantages and uniqueness of DNA cryptography methods are as follows [51]:

- The speed increases by 100 times or more when the DNA technique is used.
- The large scale of the parallelism speeds up the process of encryption and decryption using DNA techniques.
- To store 1 bit of information, 10^{12} cubic nanometres of area are required in conventional storage media, whereas DNA requires 1 cubic nanometre of area, compacting the volume of storage required to store 1 bit of information.
- Less power is required for computing DNA operations.
- Not too much time is required to manage a DNA password. It can manage everything from the encryption of data to the safe storage of data. It can also be used in information hiding, digital signature, identity authentication, etc.
- It is difficult to determine the message sequence while intercepting the ciphertext. This will assure the level of security.

Because of the above-mentioned characteristics, DNA computing is one of the most successful techniques in image encryption. But it has some drawbacks in terms of image

encryption, and these drawbacks pose challenges in image encryption. Some of the challenges are as follows [51]:

- It does not have any universal problem-solving property.
- In some models, a lot of steps for DNA operation are required. Also, more human power is required, which reduces its potential in large-scale applications.
- The DNA processing is not an automotive process; it requires human reprocessing.
- For designing a DNA password, the direct application of DNA calculation is not possible.
- The DNA password is difficult to decipher.
- Encryption and decryption of DNA password in a well-equipped biological laboratory entails huge costs.

So it is not secure to encrypt images using DNA alone. Hence, to enhance security, chaotic maps are combined with DNA cryptography to solve the hidden security problems.

4.2 Related Works

Compared to the DNA addition operation and the DNA subtraction operation, the DNA-XOR operation performs better diffusion in image encryption. Another advantage of using the DNA-XOR operation in image encryption is that the same DNA-XOR operation is used in both image encryption and decryption; hence it reduces the cost of hardware. But in the case of the DNA addition and DNA subtraction based pixel diffusion operation, DNA addition is used in image encryption whereas DNA subtraction is used in image decryption; hence, this increases the cost of hardware. The works related to the DNA-XOR operation–based chaotic image encryption are as follows.

In 2013, Zhang et al. [20] presented a novel image encryption algorithm that uses the simple theory of pseudo-DNA sequence operation to encrypt images. This algorithm combines hyper-chaotic maps, image fusion operation and DNA-XOR operation to encrypt images. In 2014, Zhang et al. [21] presented the cryptanalysis of encryption technique proposed in [20]. In [20], the authors claimed that the secret key space is large enough to resist exhaustive attack and that the key image can be periodically changed. However, in [21], the authors cryptanalysed the algorithm of [20] and found that two chaotic key-streams determined by the five keys remained unchanged for different image encryption processes. The two chaotic key-streams are recognised by choosing $4\,mn/3+1$ plaintext images. The changed period with respect to the key image should be only one. In 2014, Xie et al. [22] re-evaluated the security in [20] and found that the encryption algorithm in [20] can be broken by chosen-plaintext attack by choosing less than $\log_2(4\,mn)/2+1$ plain images.

Secret keys that are strongly related to the original image will be highly resistant to chosen-ciphertext attacks and chosen-plaintext attacks. To meet this criterion, in 2015, Guesmi et al. [23] proposed a chaos-based image encryption technique using a SHA-2 and DNA-XOR operation. In this technique, the DNA-XOR operation is used to scramble the pixels of the R, G, B components of the image. This technique not only improves information entropy but also resists various common attacks. In 2017, Enayatifar et al. [24] proposed a synchronous permutation–diffusion-based image encryption technique where

permutation and diffusion operations are performed at the same time to make the encryption process faster. Moreover, in this technique, the initial values of chaotic map required for encryption are not used directly to resist against the chosen-ciphertext attack and chosen-plaintext attack. In this technique, a 3-D logistic map is used to perform synchronous permutation and diffusion operation. Moreover, in this technique, the DNA operation is logically combined with a 3-D logistic map to increase security.

In 2017, Wu et al. [25] proposed another image encryption technique that uses the spatiotemporal chaotic system, one-way coupled-map lattices (OCML) and a DNA sequence operation such as DNA-XOR operation. The proposed technique first performs a DNA encoding operation on the individual R, G, B components. Then, it performs the DNA-XOR operation between them twice. Third, it performs a DNA decoding operation on the resultant gray-level images. Finally, the diffusion operation is performed by using a key stream generated by OCML. The proposed lossless colour image encryption technique is robust against histogram equalisation, contrast adjustment, noise addition, JPEG compression, cropping, etc.

To get the advantage of DNA-XOR operation, two image encryption schemes are described in this chapter where the DNA-XOR operation–based computing technique is used along with multiple chaotic maps to perform image encryption.

A large number of steps are required for processing DNA operation, and this may result in a slow process. For faster processing of DNA encryption, a parallel sub-image operation-based encryption is described in this chapter. The parallel sub-image encryption operation reduces the total encryption and decryption time as well as confusing attackers.

Over the past several years, many single-image encryption schemes have been proposed using the combination of DNA and chaotic maps. Single-image encryption schemes are suitable for the encryption of single images. However, for the encryption of more than one image (in this case, three images), the use of a single-image encryption scheme is not efficient because it encrypts one image at a time and hence is lacking in terms of encryption efficiency. This is a big challenge in today's digital image transmission system, where multiple images are transmitted simultaneously. Hence, to overcome all the above problems, a triple image encryption technique is described in this chapter that uses a DNA sequence operation along with multiple chaotic maps to encrypt three images simultaneously. Apart from that, the triple image encryption scheme uses parallel sub-image-based encryption operation in each of the three images to reduce the total simulation time. Both schemes resist all known common attacks, including known-plaintext attack and chosen-plaintext attacks, as well as producing larger key space and higher key sensitivity.

4.3 Preliminaries

4.3.1 DNA Sequence Operations

A DNA sequence is composed of four nucleic acid bases: Adenine (A), Guanine (G), Cytosine (C) and Thymine (T). The operations on these four DNA bases are called *DNA sequence operations*. The DNA sequence operations are equivalent to traditional binary sequence operations. All operations are performed at binary level where DNA bases are converted into corresponding binary form.

The DNA sequence operations that are used in the proposed image encryption schemes are (1) DNA encoding operation, (2) DNA decoding operation, and (3) DNA-XOR operation, which are explained below.

4.3.1.1 DNA Encoding Operation

The operation of encoding the binary bits (pair-of-bits) into DNA bases is called the *DNA encoding operation*. Out of four DNA bases (*A*, *G*, *C*, and *T*), the DNA base pair *A* and *T* are complementary to each other. Similarly, the DNA base pair *G* and *C* are complementary to each other.

In a binary system, the binary bits 0 and 1 are complementary to each other, and hence the binary bit-pairs 00 and 11, 01 and 10 are also complementary to each other. If the four DNA bases *A*, *G*, *C*, and *T* are used to encode the four binary bit-pairs 00, 01, 10, and 11, then there exists a total of $4! = 4 \times 3 \times 2 \times 1 = 24$ encoding schemes (or *encoding rules*) to encode these four binary bit-pairs. But, out of these 24 DNA encoding rules, only eight of them satisfy the Watson-Crick complementary rule [26]. Table 4.1 shows the eight rules of the DNA encoding operation.

4.3.1.2 DNA Decoding Operation

The operation of decoding the DNA bases by using eight DNA decoding rules is called the *DNA decoding operation*. The DNA decoding operation is same as the DNA encoding operation. The only difference is that the DNA encoding operation converts the binary values into DNA bases, whereas the DNA decoding operation converts the DNA bases into binary values. Table 4.1 shows the eight rules of the DNA decoding operation.

4.3.1.3 DNA-XOR Operation

The operation of XORing (exclusive-OR operation) the two DNA sequences is called the *DNA-XOR operation*. The DNA-XOR operation is performed as a traditional binary-XOR operation. For each of the eight DNA encoding/decoding rules, there exists a separate DNA-XOR rule (eight different DNA-XOR rules in total), which is shown in Table 4.2. In all the rules of DNA-XOR operation, any one base in every row or column is unique. That means the XOR operation results are unique.

TABLE 4.1

Eight Rules of DNA Encoding and Decoding Operation

	DNA Encoding Operation				DNA Decoding Operation			
Rule - 1	00 → (A)	11 → (T)	10 → (C)	01 → (G)	A → (00)	T → (11)	C → (10)	G → (01)
Rule - 2	00 → (A)	11 → (T)	01 → (C)	10 → (G)	A → (00)	T → (11)	C → (01)	G → (10)
Rule - 3	11 → (A)	00 → (T)	10 → (C)	01 → (G)	A → (11)	T → (00)	C → (10)	G → (01)
Rule - 4	11 → (A)	00 → (T)	01 → (C)	10 → (G)	A → (11)	T → (00)	C → (01)	G → (10)
Rule - 5	10 → (A)	01 → (T)	00 → (C)	11 → (G)	A → (10)	T → (01)	C → (00)	G → (11)
Rule - 6	01 → (A)	10 → (T)	00 → (C)	11 → (G)	A → (01)	T → (10)	C → (00)	G → (11)
Rule - 7	10 → (A)	01 → (T)	11 → (C)	00 → (G)	A → (10)	T → (01)	C → (11)	G → (00)
Rule - 8	01 → (A)	10 → (T)	11 → (C)	00 → (G)	A → (01)	T → (10)	C → (11)	G → (00)

TABLE 4.2
Eight Rules of DNA-XOR Operation

	Rule-1				Rule-2				Rule-3				Rule-4			
DNA-XOR (⊕)	**A**	**T**	**C**	**G**	**A**	**T**	**C**	**G**	**A**	**T**	**C**	**G**	**A**	**T**	**C**	**G**
A	A	T	C	G	A	T	C	G	T	A	G	C	T	A	G	C
T	T	A	G	C	T	A	G	C	A	T	C	G	A	T	C	G
C	C	G	A	T	C	G	A	T	G	C	T	A	G	C	T	A
G	G	C	T	A	G	C	T	A	C	G	A	T	C	G	A	T
DNA-XOR (⊕)	**Rule-5**				**Rule-6**				**Rule-7**				**Rule-8**			
A	C	G	A	T	C	G	A	T	G	C	T	A	G	C	T	A
T	G	C	T	A	G	C	T	A	C	G	A	T	C	G	A	T
C	A	T	C	G	A	T	C	G	T	A	G	C	T	A	G	C
G	T	A	G	C	T	A	G	C	A	T	C	G	A	T	C	G

4.3.2 Chaotic Maps

A chaotic map is a map that exhibits some sort of chaotic behaviour. A chaotic map may be discrete time or continuous time. The lists of chaotic maps used in the proposed encryption schemes are as follows.

4.3.2.1 Tinkerbell Map (TM)

The Tinkerbell chaotic map [27] is a 2-D discrete time system. The Tinkerbell chaotic map is described as

$$\begin{cases} x_{i+1} = x_i^2 - y_i^2 + ax_i + by_i \\ y_{i+1} = 2x_iy_i + cx_i + dy_i \end{cases} \tag{4.1}$$

In Equation 4.1, the initial values of the Tinkerbell map are represented by x and y. Similarly, the system parameters of the Tinkerbell map are represented by a, b, c, d. To exhibit the best chaotic behaviour in the Tinkerbell map, the system parameter values are commonly taken as $a = 0.9, b = -0.6013, c = 2.0, d = 0.5$.

4.3.2.2 Zaslavskii Map (ZM)

The Zaslavskii map [28] is a 2-d discrete time system. It exhibits deterministic dynamic behaviour, which is an integral part of the data encryption algorithms and is defined as:

$$\begin{cases} p_{i+1} = mod\left(p_i + v\left(1 + \mu q_i\right) + \epsilon\, v\mu cos\left(2\pi p_i\right), 1\right) \\ q_{i+1} = e^{-r}\left(q^i + cos\left(2\pi p_i\right)\right) \end{cases} \tag{4.2}$$

and

$$\mu = {\left(1 - e^{-r}\right)} \big/ {r} \tag{4.3}$$

In the above Equations 4.2 and 4.3, p and q are the initial values and r, v, ϵ are the system parameters of the Zaslavskii map. To exhibit the best chaotic behaviour in the Zaslavskii map, the system parameter values are commonly taken as $r = 3.0, v = \frac{400}{3}, \epsilon = 0.3$.

4.3.2.3 Cross-Chaotic Map (CCM)

To reduce complex calculations and to improve security, the Cross-chaotic map [29] is used. This is a 2-D map that is cross-combination of two 1-D chaotic maps. It is defined as

$$\begin{cases} g_{i+1} = 1 - sh_i^2 \\ h_{i+1} = cos(tcos^{-1}g_i) \end{cases} \tag{4.4}$$

In Equation 4.4, g, h are the initial values whose ranges are in between $[-1, 1]$, and s, t are the system parameters. To exhibit the best chaotic behaviour in the Cross-chaotic map, the system parameter values are commonly taken as $s = 2, t = 6$.

4.3.2.4 PWLCM System

PWLCM systems are widely used in image encryptions because they out-perform the Logistic map and other chaotic maps [30, 31]. The equation for PWLCM is

$$p_{n+1} = \begin{cases} \frac{p_n}{\delta} & if\ 0 \le p_n < \delta \\ \frac{p_n - \delta}{0.5 - \delta} & if\ \delta \le p_n < 0.5 \\ 1 - p_n & if\ 0.5 \le p_n < 1 \end{cases} \tag{4.5}$$

In Equation 4.5, p is the initial value and δ is the system parameter lies in the range of (0, 0.5).

Apart from the PWLCM system, a number of other famous 1-d chaotic maps are used in image encryptions, which are as follows.

Logistic Map: The Logistic map [32] is one of the simplest chaotic systems and is defined as

$$x_{n+1} = \mu x_n (1 - x_n) \tag{4.6}$$

In Equation 4.6, x is the initial value, and μ is the system parameter. In the Logistic map, to exhibit the best chaotic behaviour, the system parameter value is taken in the range of $\mu \in (3.5699456, 4)$.

Sine Map: The Sine map [33] has a similar chaotic behaviour to the Logistic map. The equation describing the Sine map is

$$x_{n+1} = a \times sin(\pi x_n) / 4 \tag{4.7}$$

In Equation 4.7, the initial value of Sine map is represented by x and $a \in (0, 4)$ is represented as system parameter.

Tent Map: The Tent map [30, 33] is a continuous time system which is defined as

$$u_{n+1} = \begin{cases} \mu u_n & \text{if } u_n < 0.5 \\ \mu(1 - u_n) & \text{otherwise} \end{cases} \tag{4.8}$$

In Equation 4.8, the initial value of Tent map is represented by u and $\mu \in (0, 2)$ is represented as system parameter.

Logistic-Sine System (LSS): This chaotic map is the combination of a Logistic map and a Sine map [33]. The equation for describing the LSS map is

$$v_{n+1} = \left(e v_n (1 - v_n) + (4 - e) \left(\frac{\sin(\pi v_n)}{4} \right) \right) mod 1 \tag{4.9}$$

In Equation 4.9, the initial value of the LSS map is represented by v and the system parameter is represented by e, whose value lies in the range of $e \in (0, 4)$. In the LSS map, the chaotic property exists in the entire range as compared to the Logistic map and the Sine map [33].

Logistic-Tent System (LTS): This chaotic map is the combination of a Logistic map and a Tent map [33]. The equation for describing the LTS map is

$$v_{n+1} = \begin{cases} \left(e v_n (1 - v_n) + (4 - e) v_n / 2 \right) mod 1 & \text{if } v_n < 0.5 \\ \left(e v_n (1 - v_n) + (4 - e)(1 - v_n) / 2 \right) mod 1 & \text{if } v_n \geq 0.5 \end{cases} \tag{4.10}$$

In Equation 4.10, the initial value of the LTS map is represented by v and the system parameter is represented by e whose value lies in the range of $e \in (0, 4)$. Similarly to the LSS map, in the LTS map, the chaotic property also exists in the entire range, in contrast to the Logistic map and the Tent map [33].

Tent-Sine System (TSS): This chaotic map is the combination of a Tent map and a Sine map [33]. The equation for describing the TSS map is

$$v_{n+1} = \begin{cases} \frac{e v_n}{2} + (4 - e)(\sin(\pi v_n) / 4) mod 1 & \text{if } v_n < 0.5 \\ \frac{e(1 - v_n)}{2} + (4 - e)(\sin(\pi v_n) / 4) mod 1 & \text{if } v_n \geq 0.5 \end{cases} \tag{4.11}$$

In Equation 4.11, the initial value of the TSS map is represented by v and the system parameter is represented by e whose value lies in the range of $e \in (0, 4)$.

Similarly to the LSS and LTS maps, in the TSS map, the chaotic property also exists in the entire range in contrast to the Sine map and Tent map [33].

4.4 Proposed Schemes

This chapter proposes two image encryption schemes using DNA sequence operations and chaotic maps:

1. A parallel sub-image encryption scheme
2. A triple image encryption scheme

In each of the schemes, three processes – the secret key generation process, encryption process, and decryption process – are required to process the whole algorithm.

4.4.1 Secret Key Generation Process

For designing any encryption algorithm, two points should be assumed.

- The algorithm is known to everyone.
- The secrecy of the algorithm exists only on the key, and hence the keys should be secret to all except the sender and the receiver.

So, the key design is much more important in any encryption algorithm. For an image encryption algorithm, the designed or generated keys should be related to the original images to resist against known-plaintext attack* and chosen-plaintext attack.† Apart from that, the keys should be highly sensitive‡ in both the encryption and decryption processes.

The secret key generation process requires the original secret keys, the original images, and the Secure Hash Algorithm SHA-256. The secret key generation processes are as follows:

Secret Key Generation Process for Parallel Sub-image Encryption Scheme: In parallel sub-image encryption scheme, four chaotic maps such as Tinkerbell map, Zaslavskii map, PWLCM system, and Cross-chaotic map are used. The initial values and system parameters of each of the chaotic maps are regarded as the secret keys of the algorithm. Apart from that, the 256-bit hash values of the SHA-256 hash algorithm§ are also taken as a secret key of the algorithm. The 64-hex values (conversion of

* In this type of attack, the attacker knows the plaintext of some portion of the ciphertext and tries to decrypt the remaining portions of the ciphertext by taking the information from the known plaintext.

† In this type of attack, the attacker tries to identify the key by knowing the chosen plaintext (selected by the attacker) and their corresponding ciphertext.

‡ Outputs are changes made by small changes in the keys.

§ SHA-256 hash algorithm is a secure one-way hash function whose aim is to provide privacy, authenticity and integrity in a form of data communication. The SHA-256 hash function generates a fixed length of 256 bits of hash values of any image size, and the main logic behind the SHA-256 hash function is high sensitivity to input images, which means, if any, small changes in the input images will lead to great changes in the output hash values.

256-bit hash values) of the original image are used to generate the initial values and system parameters of all the chaotic maps. The algorithm (Algorithm 4.1) for generating the 64-hex values is as below.

Algorithm 4.1: Generation of 64-hex Values

1. Read original image A.
2. Apply SHA-256 hash algorithm on A.
   - Generation of 256-bit hash values, $b_1, b_2, b_3, b_4, \cdots\cdots, b_{256}$
3. Combine 4-4 bits.
   - Generation of 64-hex values, $h_1, h_2, h_3, \cdots\cdots, h_{64}$

The algorithm (Algorithm 4.2) for generating the secret keys of parallel sub-image encryption scheme by using the original initial values and system parameters of the Tinkerbell map and the 64-hex values is as below.

Algorithm 4.2: Generation of Secret Keys of Tinkerbell Map

1. Generate 64-hex values, $h_1, h_2, h_3, \cdots\cdots, h_{64}$ **(Algorithm 4.1)**
2. Give initial values of Tinkerbell map, $x(1) = -0.72, y(1) = -0.64$
3. Give system parameters of Tinkerbell map, $a = 0.9, b = -0.6013, c = 2, d = 0.5$
4. Generate initial values, $x1(1), y1(1)$

$$x1(1) = x(1) - tan\Big(\big(\text{sum}(h_1 \text{ to } h_4)/10^4\big) - cos\big(\text{sum}(h_1 \text{ to } h_4)/10^4\big)\Big)/199$$

$$y1(1) = y(1) - tan\Big(\big(\text{sum}(h_5 \text{ to } h_8)/10^4\big) - cos\big(\text{sum}(h_5 \text{ to } h_8)/10^4\big)\Big)/299$$

5. Generate system parameters, $a1, b1, c1, d1$

$$a1 = a - tan\Big(\big(\text{sum}(h_9 \text{ to } h_{12})/10^4\big) - cos\big(\text{sum}(h_9 \text{ to } h_{12})/10^4\big)\Big)/399$$

$$b1 = b - tan\Big(\big(\text{sum}(h_{13} \text{ to } h_{16})/10^4\big) - cos\big(\text{sum}(h_{13} \text{ to } h_{16})/10^4\big)\Big)/499$$

$$c1 = c - tan\Big(\big(\text{sum}(h_{17} \text{ to } h_{20})/10^4\big) - cos\big(\text{sum}(h_{17} \text{ to } h_{20})/10^4\big)\Big)/599$$

$$d1 = d - tan\Big(\big(\text{sum}(h_{21} \text{ to } h_{24})/10^4\big) - cos\big(\text{sum}(h_{21} \text{ to } h_{24})/10^4\big)\Big)/699$$

In a similar way, the other secret keys of the parallel sub-image encryption scheme are generated by using the original initial values ($p(1) = -0.28, q(1) = 0.58$) and the system parameters ($r = 3, v = 400/3, E = 0.3$) of Zaslavskii map and the 64-hex values (h_{25} to h_{28}, h_{29} to h_{32}, h_{33} to h_{36}, h_{37} to h_{40}, h_{41} to h_{44}); using the original initial value ($e(1) = 0.2$) and system parameter ($ita = 0.3$) of the PWLCM system and the 64-hex values (h_{45} to h_{46}, h_{47} to h_{48}); and using the original initial values ($u(1) = 0.7199, v(1) = 0.7654$) and system parameters ($m = 6, n = 2$) of the Cross-chaotic map and the 64-hex values (h_{49} to h_{52}, h_{53} to h_{56}, h_{57} to h_{60}, h_{61} to h_{64}).

Secret Key Generation Process for the Triple Image Encryption Scheme: In the triple image encryption scheme, seven one-dimensional chaotic maps such as the Logistic

map, Sine map, Tent map, LSS map, LTS map, TSS map, and PWLCM system are used. The secret key generation process for the triple image encryption scheme is similar to the parallel sub-image encryption scheme.

The algorithm (Algorithm 4.3) for generating the secret keys of the triple image encryption scheme by using the original initial value and system parameter of the Logistic map and the 64-hex values is as below.

Algorithm 4.3: Generation of Secret Keys of Logistic Map

1. Generate 64-hex values, $h_1, h_2, h_3, \cdots\cdots, h_{64}$ (Algorithm 4.1)
2. Give initial value of Logistic map, $x(1) = 0.657$
3. Give system parameter of Logistic map, $r = 3.977$
4. Generate initial value, $x1(1)$

$$x1(1) = x(1) - tan\left(\left(sum(h_1 \text{ to } h_5)/10^4\right) - cos\left(sum(h_1 \text{ to } h_5)/10^4\right)\right)/199$$

5. Generate system parameter, $r1$

$$r1 = r - tan\left(\left(sum(h_6 \text{ to } h_{10})/10^4\right) - cos\left(sum(h_6 \text{ to } h_{10})/10^4\right)\right)/299$$

In a similar way, the other secret keys of the triple image encryption scheme are generated by using the original initial value ($y(1) = 0.997$) and system parameter ($s = 3.938$) of the Logistic-Sine system and the 64-hex values (h_{11} to h_{15}, h_{16} to h_{20}); using the original initial value ($z(1) = 0.955$) and system parameter ($t = 3.972$) of the Logistic-Tent system and the 64-hex values (h_{21} to h_{25}, h_{26} to h_{30}); using the original initial value ($q(1) = 0.983$) and system parameter ($u = 3.957$) of the Tent-Sine system and the 64-hex values (h_{31} to h_{35}, h_{36} to h_{40}); using the original initial value ($g(1) = 0.2$) and system parameter ($ita = 0.3$) of the PWLCM system and the 64-hex values (h_{41} to h_{42}, h_{43} to h_{44}); using the original initial value ($e(1) = 0.662$) and system parameter ($v = 3.999$) of the Sine map and the 64-hex values (h_{45} to h_{49}, h_{50} to h_{54}); and using the original initial value ($f(1) = 0.687$) and system parameter ($w = 3.876$) of the Tent map and the 64-hex values (h_{55} to h_{59}, h_{60} to h_{64});

4.4.2 Encryption Operation

In the first method of image encryption, a parallel sub-image encryption scheme is described. In this scheme, the image is divided into eight parts, and all the parts are operated in parallel. In each part, the DNA encoding, decoding, and XOR operations are different. This parallel operation not only decreases the total encryption and decryption time but also provides more security.

In the second method of image encryption, a triple image encryption scheme is described. In this scheme, three images are encrypted at the same time by using a single encryption algorithm, and each image is divided into eight parts to perform a parallel operation similar to the first method. The triple image encryption scheme provides better encryption efficiency as compared to the encryption of those images using a single-image encryption scheme. The difference between the first method and the second method is that the first

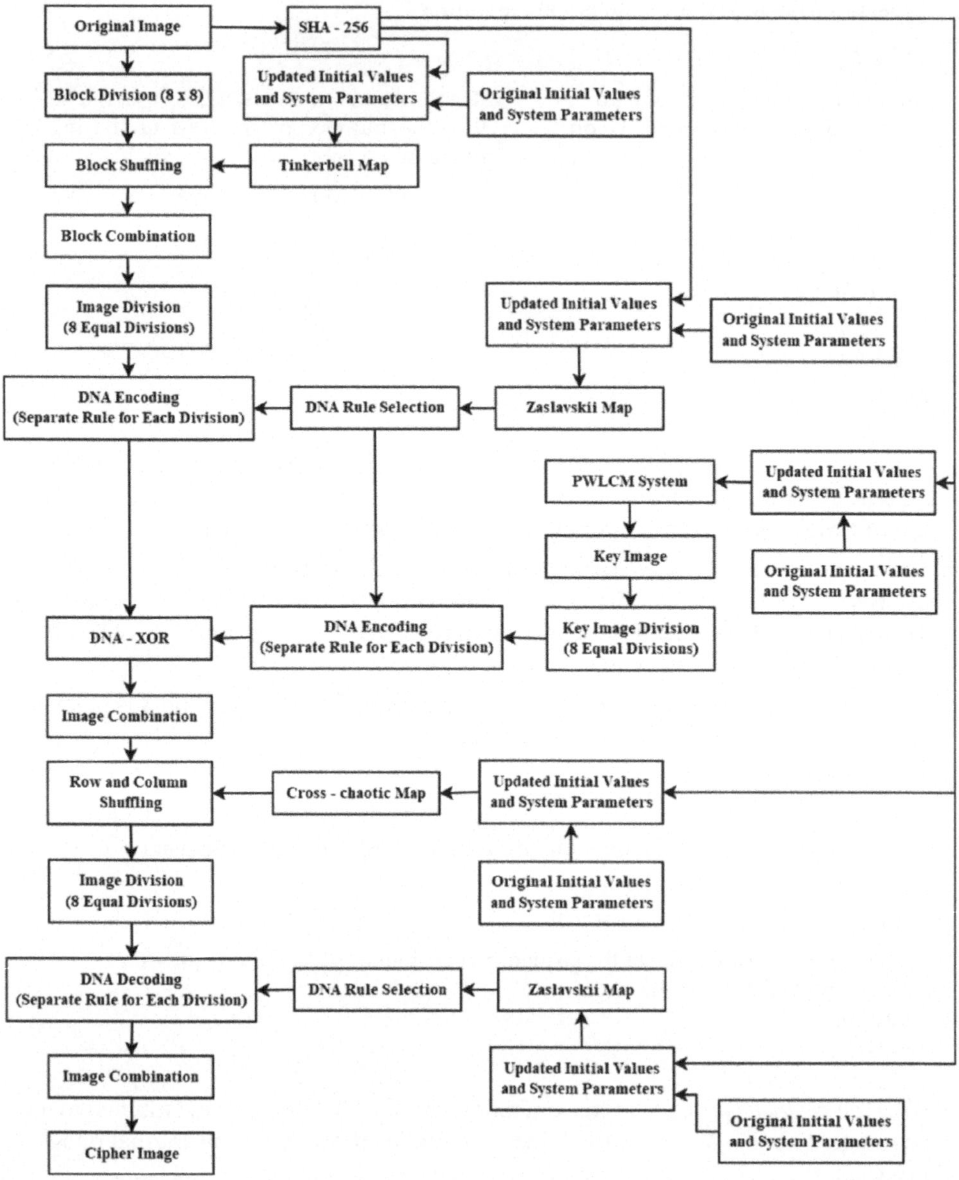

FIGURE 4.1
Encryption block diagram of the parallel sub-image encryption scheme.

method encrypts a single image at a time and the second method encrypts three images at a time.

The block diagram representing the first method is shown in Figure 4.1. The descriptions of each of the blocks of operations for Method 1 are as follows.

1. *Block division operation*: In this operation, the whole image is divided into equal sized blocks, each of size 8×8. Algorithm 4.4 shows the process of block division operation.

Algorithm 4.4: Block Division (8×8) Operation

```
m = 1, k = 1
for i = 1 to M with steps of 8 do          // M is the number of rows of the input image A
  for j = 1 to N with steps of 8 do        // N is the number of columns of the input image A
     C{k,m} = A(i to i+7, j to j+7)        // A is the input gray-scale image
  m = m + 1
  end for
  k = k + 1
  m = 1
end for
```

b. *Tinkerbell map sequence generation operation*: In this operation, the two sequences $x1$ and $y1$ of the Tinkerbell map are generated by iterating Equation 4.1, the number of times equal to the number of blocks generated in an image.

c. *Sorting and indexing Tinkerbell map sequences*: In this operation, the iterated values of the two sequences $x1$ and $y1$ of the Tinkerbell map are sorted and then indexed. The blocks are then shuffled according to the index values generated by these two sequences of Tinkerbell map.

d. *Block shuffling operation using Tinkerbell map sequences*: In this operation, the blocks are shuffled by using the index values generated by the two sequences $x1$ and $y1$ of the Tinkerbell map. Algorithm 4.5 shows the process of the block shuffling operation.

Algorithm 4.5: Block Shuffling Operation Using Tinkerbell Map Sequence

```
for i = 1 to M/8 with steps of 1 do
  for j = 1 to N/8 with steps of 1 do
  CC(i,j) = (i^th index value of the sequence x1, j^th index value of the sequence y1)
  //C is from Algorithm 4.4
  end for
end for
```

e. *Block combination operation*: In this operation, the shuffled blocks are combined to form an image. The size of the shuffled image should be the same as the original image.

f. *Block-shuffled image division (eight equal divisions) operation*: In this operation, the block-shuffled image is equally divided into eight parts. In each part, the DNA encoding, DNA-XOR, and DNA decoding operations are different.

g. *Zaslavskii map sequence generation operation*: In this operation, the two sequences $p1$ and $q1$ of Zaslavskii map are generated by iterating Equation 4.2, the number of times equal to the number of DNA encoding and decoding rules.

h. *Sorting and indexing Zaslavskii map sequences*: In this operation, the iterated values of the two sequences $p1$ and $q1$ of the Zaslavskii map are sorted and then indexed. The indexing values of $p1$ are used to select DNA encoding rules to encode both the block-shuffled image and the key image. The indexing values of $p1$ are also used to select the DNA-XOR rule for DNA-XOR operation. The indexing values of $q1$ are used to select the DNA decoding rule to decode the final encrypted image.

i. *PWLCM system sequence generation operation*: In this operation, the sequence *e*1 of the PWLCM system is generated. This sequence is used to form a key image that is the same size as the original image. Algorithm 4.6 shows the process of generating the PWLCM system sequence.

Algorithm 4.6: Generation of PWLCM System Sequence

for $i = 1$ to $M \times N \times 8 - 1$ with steps of 1 **do**
  if $e1(i) > 0.5$ // *e1* is the newly generated initial value of the PWLCM system
    $e1(i) = 1 - e1(i)$
  else
    $e1(i) = e1(i)$
  end if
  if $e1(i) \geq 0$ and $e1(i) < ita1$ // *ita1* is the newly generated system parameter of the PWLCM system
    $e1(i+1) = e1(i) / ita1$
  else if $e1(i) \geq ita1$ and $e1(i) < 0.5$
    $e1(i+1) = (e1(i) - ita1) / (0.5 - ita1)$
  else
    Display "Invalid"
  end if
end for
for $j = 1$ to $M \times N \times 8 - 1$ with steps of 1 **do**
  if $e1(j) > thr$ // *thr* is the threshold value of 0.25.
    $e1(j) = 1$
  else
    $e1(j) = 0$
  end if
end for

j. *Key image formation*: In this operation, the key image is formed by using the chaotic sequences generated by the PWLCM system. Algorithm 4.7 shows the key image formation process.

Algorithm 4.7: Key Image Formation

$z = 1$
$s = \text{zeros}(1, M \times N)$
for $j = 1$ to $M \times N \times 8 - 1$ with steps of 8 **do**
  $s(z) = 2^7 \times e1(j) + 2^6 \times e1(j+1) + 2^5 \times e1(j+2) + 2^4 \times e1(j+3)$
    $+2^3 \times e1(j+4) + 2^2 \times e1(j+5) + 2 \times e1(j+6) + e1(j+7)$
  $z = z + 1$
end for
$K = \textbf{reshape}(s, [M, N])$

k. *Key image division (eight equal divisions) operation*: In this operation, the key image is equally divided into eight parts (similar to the block-shuffled image division).

l. *DNA rule selection and encoding operation*: In this operation, each divisional part of the block-shuffled image and key image are DNA encoded by the selected DNA encoding rules. In each divisional part, the DNA encoding rules are different and hence the DNA encoding is also different. The DNA encoding rules are selected by the index values of the sequence $p1$ of the Zaslavskii map.

m. *DNA-XOR operation*: In this operation, each DNA encoded divisional images of the block-shuffled image and key image are separately DNA-XORed by the DNA-XOR rules selected by the index values of the sequence $p1$ of the Zaslavskii map.

n. *Image combination operation*: In this operation, the DNA-XORed segmented images are combined to form an image.

o. *Cross-chaotic map sequence generation operation*: In this operation, the two sequences $u1$ and $v1$ of the Cross-chaotic map are generated by iterating Equation 4.4, the number of times equal to the number of rows and columns of an image.

p. *Sorting and indexing Cross-chaotic map sequences*: In this operation, the iterated values of the two sequences $u1$ and $v1$ of the Cross-chaotic map are sorted and then indexed. The rows and columns of the DNA-XORed combined image are shuffled according to the index values generated by these two sequences of the Cross-chaotic map.

q. *Row and column shuffling operation*: In this operation, the rows and columns of DNA-XORed combined image are shuffled by using the index values generated by the two sequences $u1$ and $v1$ of the Cross-chaotic map. Algorithm 4.8 shows the process of the row-column shuffling operation.

Algorithm 4.8: Row-Column Shuffling Operation Using Cross-Chaotic Map Sequences

for $i = 1$ to M with steps of 1 **do**
  for $j = 1$ to N with steps of 1 **do**
    $O(i,j) = gkxc\left(i^{th} \text{ index value of the sequence } u1,\ j^{th} \text{ index value of the sequence } v1\right)$
  // $gkxc$ is the combined image of DNA-XOR outputs
  end for
end for

r. *Image division (eight equal divisions) operation*: In this operation, the row-column shuffled image is equally divided into eight parts similar to the key image division or the block-shuffled image division operation.

s. *DNA rule selection and decoding operation*: In this operation, each divisional part of the row-column shuffled image is DNA decoded by the selected DNA decoding rules. In each divisional part, the DNA decoding rules are different, and hence the DNA decoding is also different. The DNA decoding rules are selected by the index values generated by the sequence $q1$ of the Zaslavskii map.

t. *Image combination operation*: In this operation, separate DNA decoded images are combined to form a cipher image.

The block diagram representing the second method (triple image encryption method) is shown in Figure 4.2. The descriptions of each of the blocks of operations for Method 2 are as follows.

a. *Binary bit-plane decomposition operation*: In this operation, eight bit-planes are generated in each of the three images. The first bit-plane is formed by combining the first bit of all the pixels. Similarly, the second bit-plane is formed by combining the second bit of all the pixels, and so on until all eight bit-planes are formed.
b. *Logistic map sequence generation operation*: In this operation, a sequence $x1$ of the Logistic map is generated by iterating Equation 4.6 $(24+8)$ times. The first 24 iteration outputs of the sequence $x1$ are used to perform a bit-plane shuffling operation between the three images. The last eight iteration outputs of the sequence $x1$ are used to select the DNA-XOR rule to perform the DNA-XOR operation between the three images and the key image.
c. *Sorting and indexing Logistic map sequences*: In this operation, the first 24 iterated values and the last eight iterated values of the sequence $x1$ of Logistic map are sorted independently and then indexed. The bit-planes of the three images are shuffled according to the index values generated by the first 24 iterated values of the sequence $x1$, and the DNA-XOR rule is selected according to the index values generated by the last eight iterated values of the sequence $x1$ of the Logistic map.
d. *Bit-plane shuffling operation using Logistic map sequence*: In this operation, the bit-planes of the three images are shuffled according to the index values generated by the first 24 iterated values of the sequence $x1$ of the Logistic map.
e. *Bit-plane combination operation*: In this operation, the bit-planes are combined to form images. First, eight shuffled bit-planes are combined to form the first image; second, eight shuffled bit-planes are combined to form the second image; and last, eight shuffled bit-planes are combined to form the third image.
f. *Logistic-Sine system sequence generation operation*: In this operation, a sequence $y1$ of the Logistic-Sine system is generated by iterating Equation 4.9 $(8+8)$ times. The first eight iterated outputs and the last eight iterated outputs of the sequence $y1$ are used to select DNA encoding rules to perform the DNA encoding operation of the first bit-plane shuffled image and the key image, respectively.
g. *Sorting and indexing Logistic-Sine system sequences*: In this operation, the first and the last eight iterated values of the sequence $y1$ are sorted and then indexed. These index values are used to select DNA encoding rules to encode the first bit-plane shuffled image and the key image.
h. *Logistic-Tent system sequence generation operation*: In this operation, a sequence $z1$ of the Logistic-Tent system is generated by iterating Equation 4.10 $(8+8)$ times. The first eight iterated outputs and the last eight iterated outputs of the sequence $z1$ are used to select DNA encoding rules to perform the DNA encoding operation of the second bit-plane shuffled image and the same key image, respectively.
i. *Sorting and indexing Logistic-Tent system sequences*: In this operation, the first and the last eight iterated values of the sequence $z1$ are sorted and then indexed. These index values are used to select DNA encoding rules to encode the second bit-plane shuffled image and the key image.

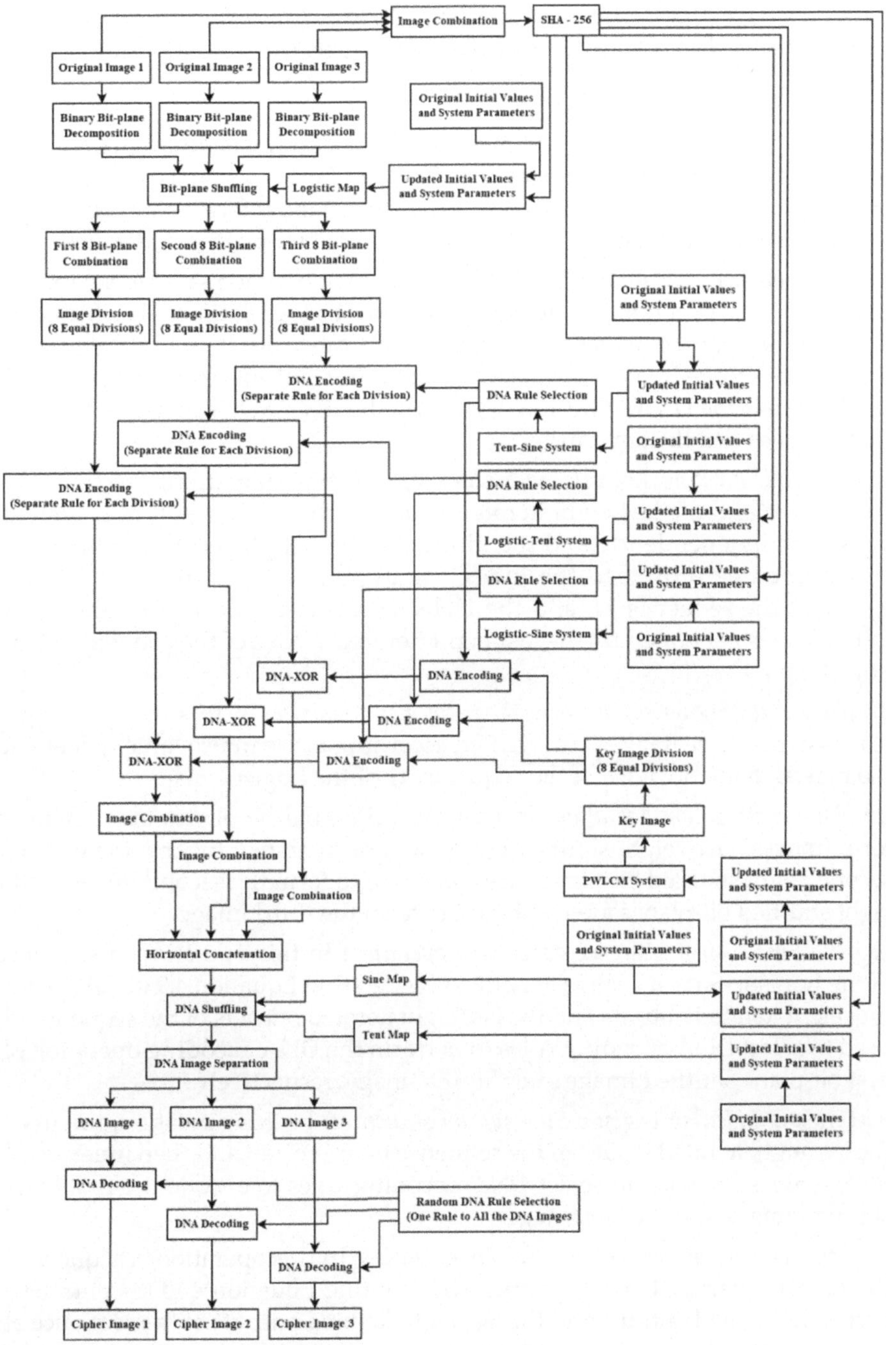

FIGURE 4.2
Encryption block diagram of triple image encryption scheme.

j. *Tent-Sine system sequence generation operation*: In this operation, a sequence $q1$ of the Tent-Sine system is generated by iterating Equation 4.11 $(8+8)$ times. The first eight iterated outputs and the last eight iterated outputs of the sequence $q1$ are used to select DNA encoding rules to perform the DNA encoding operation of the third bit-plane shuffled image and the same key image, respectively.

k. *Sorting and indexing Tent-Sine system sequences*: In this operation, the first and the last eight iterated values of the sequence $q1$ are sorted and then indexed. These index values are used to select DNA encoding rules to encode the third bit-plane shuffled image and the key image.

l. *PWLCM system sequence generation operation*: In this operation, the sequence $g1$ of the PWLCM system is generated by iterating Equation 4.5 $(M \times N \times 8)$ times. This sequence is used to form a key image that is the same size as the original image. The algorithm for generating the PWLCM system sequence is same as in Algorithm 4.6.

m. *Key image formation*: In this operation, the key image is formed by using the chaotic sequences generated by the PWLCM system. The process for the formation of the key image is same as in Algorithm 4.7.

n. *Bit-plane shuffled image and key image segmentation (or image division) operation*: In this operation, each of the three bit-plane shuffled images and the key image are segmented into eight parts.

o. *DNA rule selection and encoding operation*: In this operation, the segmented parts of each of the three images and the key image are DNA encoded by the selected DNA encoding rules. In each segmented part, the DNA encoding rules are different and hence the DNA encoding is also different. The DNA encoding rules of the first bit-plane shuffled image, the second bit-plane shuffled image, and the third bit-plane shuffled image are selected by the index values generated by the sequence $y1$ (first eight iterated values) of the Logistic-Sine system, $z1$ (first eight iterated values) of the Logistic-Tent system, and $q1$ (first eight iterated values) of the Tent-Sine system, respectively. In this operation, the same segmented key image is DNA encoded three times by using the index values $y1$ (last eight iterated values) of the Logistic-Sine system, $z1$ (last eight iterated values) of the Logistic-Tent system, and $q1$ (last eight iterated values) of the Tent-Sine system. This DNA rule selection and these DNA encoding operations are performed independently and in parallel for each of the images.

p. *DNA-XOR operation*: In this operation, the DNA encoded (after segmentation) part of the first image is DNA-XORed with the DNA encoded (after segmentation) key image (the DNA encoding rule for the key image is selected by the index values generated by the sequence $y1$ [last eight iterated values] of the Logistic-Sine system) by the DNA-XOR rule (the DNA-XOR rule is selected by the index values generated by the sequence $x1$ [last eight iterated values] of the Logistic map). Similarly, the DNA encoded part of the second image is DNA-XORed with the DNA encoded key image (the DNA encoding rule for the key image is selected by the index values generated by the sequence $z1$ [last eight iterated values] of the Logistic-Tent system) by the DNA-XOR rule (the DNA-XOR rule is selected by the index values generated by the sequence $x1$ [last eight iterated values] of the Logistic map).

The DNA encoded part of third image is DNA-XORed with the DNA encoded key image (the DNA encoding rule is selected by the index values generated by the sequence $q1$ [last eight iterated values] of the Tent-Sine system) by the DNA-XOR rule (the DNA-XOR rule is selected by the index values generated by the sequence $x1$ [last eight iterated values] of the Logistic map).

q. *Image combination operation*: In this operation, the DNA-XORed images are combined to form three images.

r. *Image concatenation operation*: In this operation, the three images are horizontally concatenated to form a large combined image.

s. *Sine map sequence generation operation*: In this operation, a sequence $e1$ of the Sine map is generated by iterating Equation 4.7, the number of times equal to the number of rows in the horizontally concatenated image. This sequence is used to perform the DNA base shuffling operation in the concatenated image.

t. *Sorting and indexing Sine map sequence*: In this operation, the iterated values of sequence $e1$ of the Sine map are sorted and then indexed. The indexing values of $e1$ are used to perform the DNA base shuffling operation in the horizontally concatenated images.

u. *Tent map sequence generation operation*: In this operation, a sequence $f1$ of the Tent map is generated by iterating Equation 4.8, the number of times equal to the number of columns in the horizontally concatenated image. This sequence is used to perform the DNA base shuffling operation in the concatenated image.

v. *Sorting and indexing Tent map sequence*: In this operation, the iterated values of sequence $f1$ of the Tent map is sorted and then indexed. The indexing values of $f1$ are used to perform the DNA base shuffling operation in the horizontally concatenated images.

w. *DNA bases shuffling operation using Sine and Tent map sequences*: In this operation, the DNA bases of the horizontally concatenated image are shuffled by using the index values generated by the sequence $e1$ of the Sine map and $f1$ of the Tent map. Algorithm 4.9 shows the process of the DNA base shuffling operation.

Algorithm 4.9: DNA Base Shuffling Operation Using Sine Map and Tent Map Sequences

for $i = 1$ to M with steps of 1 **do**
  for $j = 1$ to $3N \times 4$ with steps of 1 **do**
  $SS(i, j) = SD(i^{th}$ index value of the sequence $e1$, j^{th} index value of the sequence $f1)$
  //SD is the horizontally concatenated image
  end for
end for

x. *DNA image separation operation*: In this operation, the concatenated shuffled image is separated into three images.

y. *DNA rule selection and decoding operation*: In this operation, out of eight DNA decoding rules, any one DNA decoding rule is selected randomly to decode all the three images independently. This generates three cipher images.

4.4.3 Decryption Operation

The decryption operations of both the methods are the reverse of the encryption operations. The steps for describing the decryption operation of Method 1 (parallel sub-image encryption scheme) are as follows.

Step 1: The receiver obtains the cipher image, 256-bit hash key, original secret keys of the Tinkerbell map, Zaslavskii map, PWLCM system along with the threshold value, and Cross-chaotic map.

Step 2: In the initial stage, the cipher image is divided into eight parts, similar to image division operations in encryption.

Step 3: The two chaotic sequences $p1$ and $q1$ of the Zaslavskii map are generated and then the two sequences are sorted and indexed.

Step 4: Each of the divisional parts is DNA encoded using the DNA encoding rules. The DNA encoding rules are selected according to the index values generated by the sequence $q1$ of the Zaslavskii map.

Step 5: All the DNA encoded divisional parts are combined to form an image.

Step 6: The two sequences $u1$ and $v1$ of the Cross-chaotic map are generated and then sorted and indexed.

Step 7: A row-column de-shuffling operation is performed by using the index values generated by the sequences $u1$ and $v1$ of Cross-chaotic map. The row-column de-shuffling operation is the exact reverse of the row-column shuffling operation in encryption.

Step 8: The row-column de-shuffled image is then divided into eight parts.

Step 9: The chaotic sequence $e1$ of the PWLCM system is generated, and then, by using the PWLCM system chaotic sequence, a key image is generated, similar to Algorithm 4.6 and 4.7.

Step 10: The key image is divided into eight parts.

Step 11: Each of the divisional parts of key image is DNA encoded using the DNA encoding rules. The DNA encoding rules are selected according to the index values generated by the sequence $p1$ of the Zaslavskii map.

Step 12: The outputs of Steps 8 and 11 are DNA-XORed using the DNA-XOR rules selected by the index values generated by the sequence $p1$ of the Zaslavskii map.

Step 13: The DNA decoding operation is performed in each of the segmented parts (outputs of Step 12) by using the index values generated by the sequence $p1$ of the Zaslavskii map.

Step 14: All the DNA decoded parts are combined to get an image.

Step 15: The whole image is divided into blocks of equal size, each measuring 8×8.

Step 16: The two chaotic sequences $x1$ and $y1$ of the Tinkerbell map are generated and then sorted and indexed.

Step 17: The block de-shuffling operation is performed by using the index values generated by the sequences $x1$ and $y1$ of the Tinkerbell map. The block de-shuffling operation is the exact reverse of the block shuffling operation in encryption.

Step 18: Finally, all the blocks are combined to get a decrypted image.

In a similar way, the decryption operation of Method 2 (triple image encryption scheme) can be described.

4.5 Results and Discussions

Simulation results show the encryption and decryption outputs obtained by using Method 1 and Method 2. In Method 1, one can use any size of image, but in Method 2, all three images should be the same size. Simulation results of Methods 1 and 2 are shown in Figure 4.3. In Method 1, a "Lena" image of size 512×512 is used to perform the simulation operation. In Method 2, a group of three same-sized (256×256) images, "Cameraman," "Rice," and "Lena," are used to perform the simulation operation. Method 1 and Method 2 simulation operations are performed on a system with the following hardware configuration: 3.40GHz i7 processor, 4GB RAM, and Windows 10 operating system. MATLAB® R2012a is used as a compiling software. The simulation results from Methods 1 and 2 in Figure 4.3 show that the encrypted images are really noisy, proving that no information can be retrieved from them about their corresponding original images. The simulation results from Methods 1 and 2 in Figure 4.3 also show that the encrypted images can be properly decrypted by using the right secret keys. It can thus be concluded that the two image encryption methods provide good encryption results and are effective for encrypting gray-scale images.

4.6 Security Analysis

Security analysis describes the security of an encryption and decryption algorithm. An encryption algorithm is good for use in real-time applications, resisting all kinds of known common attacks. The explanations of each of the security analyses are as below.

FIGURE 4.3
Simulation results of parallel sub-image encryption scheme: (a) original "Lena" image, (b) encrypted "Lena" image, (c) decrypted "Lena" image; simulation results of triple image encryption scheme: (d) original "Cameraman" image, (e) encrypted "Cameraman" image, (f) decrypted "Cameraman" image, (g) original "Rice" image, (h) encrypted "Rice" image, (i) decrypted "Rice" image, (j) original "Lena" image, (k) encrypted "Lena" image, (l) decrypted "Lena" image.

4.6.1 Key space Analysis

According to Menezes [34] in his book *Handbook of Applied Cryptography*:

> "A necessary, but usually not sufficient, condition for an encryption scheme to be secure is that the key space be large enough to preclude exhaustive search."

Key space is defined as the total number of different keys used in an encryption process. In the present day of computing applications, a key any bigger than 128 bits is cryptographically secure. Hence, the key space of any encryption algorithm should be larger than 2^{128} to resist brute-force attack [35]. A large key space strengthens the encryption algorithm to resist brute-force attack for a long time.

The secret keys used in Method 1 of the image encryption operation are

a. The initial values x, y and the system parameters a,b,c,d of the Tinkerbell map; the initial values p, q and the system parameters r,v,E of the Zaslavskii map; the initial value e and the system parameter *ita* of the PWLCM system; and the initial values u, v and the system parameters m, n of the Cross-chaotic map
b. The threshold value of the PWLCM system
c. The 256-bit hash value

The secret keys used in Method 2 of the image encryption operation are

a. The initial value x and the system parameter r of the Logistic map, the initial value y and the system parameter s of the Logistic-Sine system, the initial value z and the system parameter t of the Logistic-Tent system, the initial value q and the system parameter u of the Tent-Sine system, the initial value g and the system parameter *ita* of the PWLCM system, the initial value e and the system parameter v of the Sine map, and the initial value f and the system parameter w of the Tent map
b. The threshold value of the PWLCM system
c. The 256-bit hash value
d. A random value for the DNA decoding rule selection.

Basically, a 64-bit double precision format is used for initialising and updating the initial values and system parameters of all the chaotic maps of both the methods. For the 64-bit double precision format, a precision of 10^{-15} is suggested by the IEEE floating point standard [36]. So, a precision of 10^{-15} is used to define the initial values and system parameters of all the chaotic maps of both the methods. Furthermore, the security of resisting the best attack in the SHA-256 hash function is 2^{128}.

Therefore, the total key space in Method 1 of the encryption scheme is

$$\left(10^{15}\times10^{15}\times10^{15}\times10^{15}\times10^{15}\times10^{15}\right)\times\left(10^{15}\times10^{15}\times10^{15}\times10^{15}\times10^{15}\right)\times\left(10^{15}\times10^{15}\times10^{2}\right)$$

$$\times\left(10^{15}\times10^{15}\times10^{15}\times10^{15}\right)\times2^{128}=10^{257}\times2^{128}=1.665\times2^{853}\times2^{128}=1.665\times2^{981}$$

Similarly, the total key space in Method 2 of the encryption scheme is

$$\left(10^{15}\times10^{15}\right)\times\left(10^{15}\times10^{15}\right)\times\left(10^{15}\times10^{15}\right)\times\left(10^{15}\times10^{15}\right)\times\left(10^{15}\times10^{15}\times10^{2}\right)$$

$$\times\left(10^{15}\times10^{15}\right)\times\left(10^{15}\times10^{15}\right)\times2^{128}\times10^{1}=10^{213}\times2^{128}=1.4855\times2^{707}\times2^{128}=1.4855\times2^{835}$$

This much key space for both the methods is much larger than 2^{128} to resist brute-force attack very efficiently. These larger key spaces strengthen the algorithms to resist brute-force attacks for a long time.

4.6.2 Statistical Attack Analysis

In 1949, C. E. Shannon published the paper "Communication Theory of Secrecy System" in the *Bell System Technical Journal* [37]. In this paper, one important point marked out is that statistical analysis is an important factor in determining the strength of an encryption system, and if it cannot resist statistical attack, any encryption system can be easily broken [38].

The two tests that are commonly used to determine resistance against statistical attack are *histogram analysis* and *correlation analysis.*

4.6.2.1 Histogram Analysis

A *histogram* is a graph that shows the frequency distribution of all the gray-level pixel intensity values of an image. Histogram analysis analyses the randomness of pixels in images simply by observing the pixel intensity levels in each gray level. More uniformity in gray-level pixel intensity values implies more randomness of pixels.

For a good encryption algorithm, the histograms of encrypted images should be

- Different than the histograms of original images.
- Uniform in all the gray-level pixel intensity values.

Figure 4.4 shows the histogram results of the Method 1 and Method 2 encryption systems. In this figure, the histograms of all the original images and the histograms of all the encrypted images are totally different to each other, and the histograms of all the encrypted images in both the methods are uniformly distributed in all the gray-level pixel intensity values. This uniformity of all the gray-level pixel intensity values in the histograms of encrypted images proves that a greater randomness occurs in the pixels of encrypted images. Hence, both the methods strongly resist statistical attack.

Histogram analysis only shows the pictorial view of uniformity of gray-level pixel intensity values in the encrypted images and does not show the amount of uniformity of the gray-level pixel intensity values in the histograms of the encrypted images. So quantitative analysis is needed to test uniformity in the gray-level pixel intensity values in the histograms of images. Variance is one such analysis technique for measuring uniformity in the gray-level pixel intensity values in the histograms of images.

Variance analysis measures the uniformity in the gray-level pixel intensity values in the images. The lower the variance, the higher the uniformity. Inversely, the higher the variance, the lower the uniformity [39]. The mathematical expression for calculating the variance is as follows:

$$variance(X) = \frac{1}{n^2}\sum_{i=1}^{n}\sum_{j=1}^{n}\frac{(x_i - x_j)^2}{2} \tag{4.12}$$

where:

n is the number of gray-scale values

x_i and x_j are the number of pixels for a particular gray-scale values i and j, respectively

X is the vector of all x_i's and x_j's

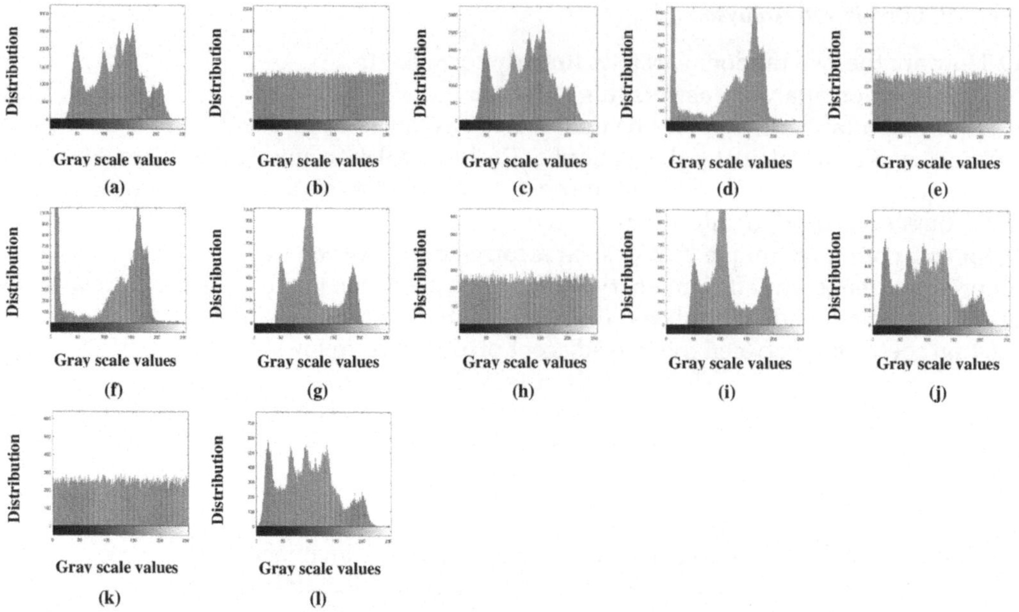

FIGURE 4.4
Histogram results of "Lena" image using parallel sub-image encryption scheme: (a) original "Lena" image, (b) encrypted "Lena" image, (c) decrypted "Lena" image; histogram results of triple image encryption scheme: (d) original "Cameraman" image, (e) encrypted "Cameraman" image, (f) decrypted "Cameraman" image, (g) original "Rice" image, (h) encrypted "Rice" image, (i) decrypted "Rice" image, (j) original "Lena" image, (k) encrypted "Lena" image, (l) decrypted "Lena" image.

Table 4.3 presents the results of variances of different images by using both the methods. Table 4.3 shows that the variances in the original images are much larger than the variances in encrypted images. That means a larger nonuniformity occurs in the histograms of original images and a larger uniformity occurs in the histograms of encrypted images. So, a larger uniformity proves a larger randomness of pixels in the encrypted images. A larger randomness of pixels proves a stronger security in the encryption systems.

TABLE 4.3
Results of Histogram Variances Using Methods 1 and 2

Image Encryption Schemes		Original Images	Encrypted Images
Method 1 (parallel sub-image encryption)	Cameraman (256×256)	1.1097e+05	261.2578
	Lena (512×512)	6.3340e+05	1.2132e+03
	Baboon (512×512)	7.4943e+05	1.2824e+03
	Peppers (512×512)	4.8066e+05	1.3273e+03
Method 2 (triple image encryption)	Cameraman (256×256)	1.1097e+05	265.2813
	Rice (256×256)	9.6312e+04	251.3047
	Lena (256×256)	3.0588e+04	278.5703
Ref. [47]	Lena	—	5554.8293
Ref. [48]	Lena	—	5335.8309

4.6.2.2 Correlation Analysis

Q. How are the two random variables linearly correlated?

A. Correlation analysis answers this question.

Let the two random variables be denoted as two adjacent pixels of an image. The correlation coefficient calculates the amount of linear correlation among the adjacent pixels in an image. In a meaningful original image, two adjacent pixels are highly correlated in all three directions (horizontal, vertical, and diagonal).

An encryption technique breaks such a correlation between the two adjacent pixels in an original image. An efficient encryption technique is one that breaks the correlation of almost all pairs of adjacent pixels. The mathematical expressions for calculating the correlation coefficient between the two adjacent pixels are as follows:

$$r_{xy} = \frac{cov(x,y)}{\sqrt{D(x)D(y)}} \tag{4.13}$$

where:

x and y are the two adjacent gray-level pixel values of an image

r_{xy} is the correlation coefficient of two adjacent pixels x and y

$cov(x,y)$ denotes the covariance of x and y, which is expressed as

$$cov(x,y) = \frac{1}{N}\sum_{i=1}^{N}(x_i - E(x))(y_i - E(y)) \tag{4.14}$$

$D(x)$ and $D(y)$ denote the variances of x and y respectively, which are expressed as

$$D(x) = \frac{1}{N}\sum_{i=1}^{N}(x_i - E(x))^2 \text{ and } D(y) = \frac{1}{N}\sum_{i=1}^{N}(y_i - E(y))^2 \tag{4.15}$$

$E(x)$ and $E(y)$ denote the expectations or the mean values of x and y respectively, which are expressed as

$$E(x) = \frac{1}{N}\sum_{i=1}^{N}x_i,\ E(y) = \frac{1}{N}\sum_{i=1}^{N}y_i \tag{4.16}$$

The value for correlation coefficient r_{xy} is in between −1 and 1. If it is greater than 0, it indicates positive correlation; if it is less than 0, it indicates negative correlation; and if it is equal to 0, it indicates zero correlation or noncorrelation of adjacent pixels [40]. For an original image, the correlation coefficients are between +1 and −1, but for a good encryption algorithm to withstand statistical effects, the correlation coefficient value for encrypted images should be very close to 0. Table 4.4 shows the correlation coefficient results of original images and encrypted images obtained by using Methods 1 and 2. Both the methods give correlation coefficient values nearer to +1 in original images and correlation coefficient values nearer to 0 in encrypted images in all the three directions. This proves that a strong correlation exists in the pixels of original images and a weak correlation exists in the pixels of encrypted images.

TABLE 4.4

Results of Correlation Coefficient Using Methods 1 and 2

Image Encryption Schemes		Original Images Horz.	Original Images Vert.	Original Images Diag.	Encrypted Images Horz.	Encrypted Images Vert.	Encrypted Images Diag.
Method 1 (parallel sub-image encryption)	Cameraman (256×256)	0.9319	0.9609	0.9077	0.0019	0.0020	0.0066
	Lena (512×512)	0.9749	0.9837	0.9592	−0.0012	0.0026	−0.0023
	Baboon (512×512)	0.8615	0.7522	0.7285	−0.0036	−0.0038	0.0015
	Peppers (512×512)	0.9754	0.9799	0.9615	−0.0094	−0.0082	0.0026
Method 2 (triple image encryption)	Cameraman (256×256)	0.9319	0.9609	0.9077	−0.0003	−0.0034	−0.0081
	Rice (256×256)	0.9259	0.9422	0.8925	−0.0001	−0.0011	−0.0064
	Lena (256×256)	0.9382	0.9685	0.9209	−0.0057	−0.0041	0.0049
Ref. [15]	Lena (256×256)	—	—	—	0.0058	0.0061	0.0059
Ref. [14]	Lena (256×256)	—	—	—	0.0030	−0.0024	−0.0034

Horz., Horizontal; *Vert.*, Vertical; *Diag.*, Diagonal.

Figure 4.5 shows the correlation distribution results of the "Lena" image by using Method 1, and Figure 4.6 shows the correlation distribution results of the group of three images by using Method 2. Both the figures show that the two adjacent pixels in all the three directions in the original images are highly correlated, whereas in encrypted images, the adjacent pixels in all three directions are scattered throughout the entire gray-level space and hence are lightly correlated or zero correlated. This proves a stronger resistivity to statistical attack in both the methods.

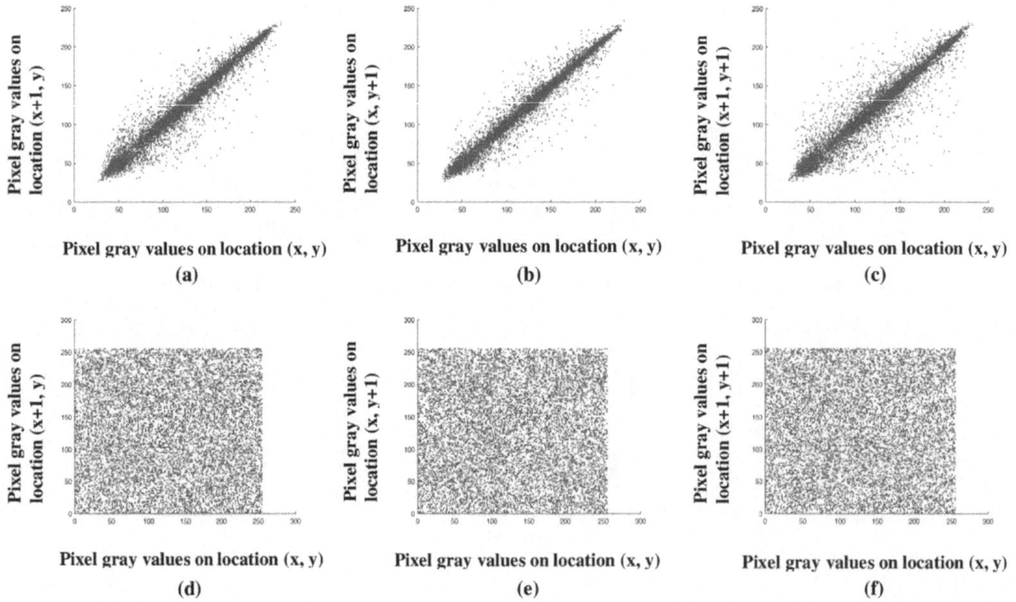

FIGURE 4.5

Correlation distribution results of "Lena" image using parallel sub-image encryption technique: (a) horizontal distribution of original image, (b) vertical distribution of original image, (c) diagonal distribution of original image, (d) horizontal distribution of encrypted image, (e) vertical distribution of encrypted image, (f) diagonal distribution of encrypted image.

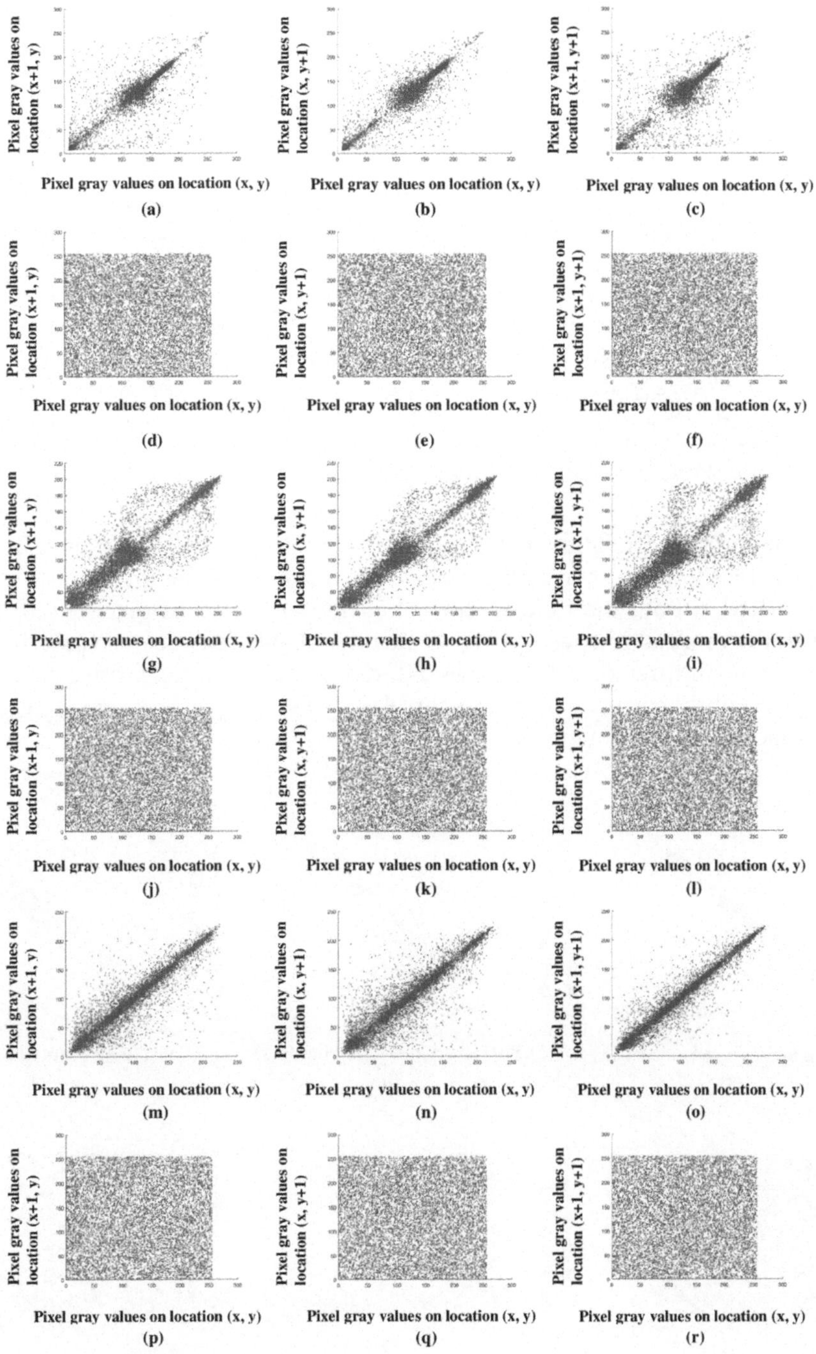

FIGURE 4.6
Correlation distribution results of "Cameraman," "Rice," and "Lena" images using triple image encryption technique: (a), (g), (m) horizontal distribution of original "Cameraman," "Rice," and "Lena" images, respectively; (b), (h), (n) vertical distribution of original "Cameraman," "Rice," and "Lena" images, respectively; (c), (i), (o) diagonal distribution of original "Cameraman," "Rice," and "Lena" images, respectively; (d), (j), (p) horizontal distribution of encrypted "Cameraman," "Rice," and "Lena" images, respectively; (e), (k), (q) vertical distribution of encrypted "Cameraman," "Rice," and "Lena" images, respectively; (f), (l), (r) diagonal distribution of encrypted "Cameraman," "Rice," and "Lena" images, respectively.

4.6.3 Differential Attack Analysis

Attackers always try to capture some useful information from the encrypted image, so as to determine a relationship between the original image and the corresponding encrypted image. Hence, encryption algorithms try to produce a large difference between the original image and the corresponding encrypted image so that no information will be available to the attacker.

A differential attack is one such attack, used by attackers to find a relationship between the original image and the encrypted image. In this type of attack, attackers simply change a random pixel value in an original image and then apply the encryption algorithm to this changed image to get a new encrypted image. Finally, by comparing the two encrypted images (the original encrypted image and the new encrypted image), the attackers work out the relationship between the original image and the encrypted image [23].

To measure such a relationship (i.e., to measure the effectiveness of a differential attack), two security criteria, NPCR (Numbers of Pixel Changing Rate) and UACI (Unified Average Changing Intensity), are used.

4.6.3.1 Numbers of Pixel Changing Rate (NPCR)

NPCR measures the rate of change of pixel values in an encrypted image by changing just one pixel value in the original image. The closer it gets to 100%, the more it becomes sensitive to plaintext and the more it resists plaintext attack [41].

The formula for calculating NPCR is

$$NPCR = \left(\frac{\left(\sum_{i=1}^{M} \sum_{j=1}^{N} D(i,j) \right)}{M \times N} \right) \times 100\% \tag{4.17}$$

where:

$M \times N$ is the size of the image matrix

$D(i,j)$ Represents the equivalency of two encrypted images which is given by

$$D(i,j) = \begin{cases} 0 & if\ C_1(i,j) = C_2(i,j) \\ 1 & if\ C_1(i,j) \neq C_2(i,j) \end{cases} \tag{4.18}$$

where:

C_1 is the original encrypted image

C_2 is the changed encrypted image, where one pixel in the original image has been changed.

For an 8-bit gray-scale image, the expected estimated NPCR value of a good encryption algorithm is around 99.6094% [42]

4.6.3.2 Unified Average Changing Intensity (UACI)

UACI measures the average change in intensity between the original image and the corresponding encrypted image. The larger UACI value strongly resists differential attack [43].

The formula for calculating UACI is

$$UACI = \left(\frac{\left(\sum_{i=1}^{M} \sum_{j=1}^{N} \left| C_1(i,j) - C_2(i,j) \right| \right)}{255 \times M \times N} \right) \times 100\% \qquad (4.19)$$

where:

$M \times N$ is the size of the image matrix

C_1 is the original encrypted image, and C_2 is the changed encrypted image, where one pixel value in the original image has been changed.

For an 8-bit gray-scale image, the expected estimated UACI value of a good encryption algorithm is around 33.4635% [42].

Table 4.5 shows the minimum, maximum, and average NPCR and UACI results obtained by using both the methods. For calculating minimum, maximum, and average NPCR and UACI values, both the methods are repeated 100 times by randomly changing the pixel value. In Table 4.5, the average NPCR and UACI values are greater than the expected estimated NPCR and UACI values. This shows that using both methods leads to a strong resistivity to differential attack.

4.6.4 Information Entropy Analysis

In 1948, Claude Shannon first introduced the concept of information entropy in his paper "A Mathematical Theory of Communication" [44]. In a system, the information entropy is defined as expressing the degree of uncertainty of the system information. If the system is considered as an image, then in an image, the information entropy is defined as expressing the degree of uncertainty or degree of randomness of the image information [45].

In common parlance, information entropy simply measures the distribution of gray-level pixel values in an image. The more uniform the distribution of gray-level pixel values, the

TABLE 4.5

Results of NPCR and UACI Using Methods 1 and 2

		Encrypted Images					
		NPCR (%)			UACI (%)		
Image Encryption Schemes		**Min.**	**Max.**	**Avg.**	**Min.**	**Max.**	**Avg.**
Method 1	Cameraman (256×256)	99.5873	99.6426	99.6123	33.4324	33.5178	33.4899
	Lena (512×512)	99.5994	99.6528	99.6279	33.4570	33.5596	33.5160
	Baboon (512×512)	99.5888	99.6419	99.6188	33.4426	33.5108	33.4729
	Peppers (512×512)	99.5937	99.6498	99.6261	33.4528	33.5427	33.4957
Method 2	Cameraman (512×512)	99.5878	99.6421	99.6100	33.4176	33.5046	33.4639
	Lena (512×512)	99.5845	99.6537	99.6123	33.4455	33.5567	33.4927
	Rice (512×512)	99.5736	99.6386	99.6099	33.4234	33.5098	33.4671
Ref. [47] (1 round)		—	—	0.4222	—	—	0.1365
Ref. [48] (1 round)		—	—	99.5838	—	—	17.0035

more the information entropy [45, 46]. The formula for describing information entropy is as follows:

$$H(m) = -\sum_{i=0}^{2^L-1} p(m_i)\log_2 p(m_i) \tag{4.20}$$

where:

- m represents the information source
- $H(m)$ represents the information entropy for the information source m
- $p(m_i)$ represents the probability of each of the symbols m_i
- $\log_2$ represents the logarithm in base 2. The base 2 logarithm is basically used because the pixel values are represented in bits.
- L is the number of bits to represent each of the gray-level pixel values. Since the gray-level pixel values are between 0 and 255, the number of bits required to represent each of the pixel values is 8.

$$\therefore 2^L - 1 = 2^8 - 1 = 256 - 1 = 255 \tag{4.21}$$

For an ideal random gray-scale image, the information entropy value is 8. This will be explained below.

For an ideal random gray-scale image, the probabilities of each of the symbols m_i are equal. Since there are 256 gray levels, the probabilities of each of the symbols m_i are 1/256.

$$\begin{aligned}\therefore H(m) &= -\sum_{i=0}^{255} p(m_i)\log_2 p(m_i) \\ &= -\sum_{i=0}^{255} \tfrac{1}{256}\log_2 \tfrac{1}{256} \\ &= -256 \times \tfrac{1}{256}\log_2 \tfrac{1}{2^8} \\ &= -\log_2 2^{-8} = -(-8)\log_2 2 = 8\end{aligned} \tag{4.22}$$

The encrypted image after encryption should behave as an ideal random image. That mean the information entropy value of an encrypted image should be close to the value 8. The more it becomes close to 8, the less is the leakage of information. Table 4.6 presents the information entropy results of various images obtained by using both the methods. Table 4.6 shows that all the encrypted images have information entropy values by using both the methods are very nearer to the value of 8. This proves that a greater randomness occurs in the pixels of encrypted images obtained by using both the methods, and hence the leakage of information is reduced. This shows that a stronger resistivity to information entropy attack is obtained when both the methods are used.

4.6.5 Key Sensitivity Analysis

Key sensitivity measures the sensitivity of the key in the algorithm. That means that it simply measures the change in output made by small changes in the key. NPCR and UACI are

TABLE 4.6
Results of Information Entropy Using Methods 1 and 2

Image Encryption Schemes		Original Images	Encrypted Images
Method 1 (parallel sub-image encryption)	Cameraman (256×256)	7.0097	7.9971
	Lena (512×512)	7.4451	7.9992
	Baboon (512×512)	7.3583	7.9991
	Peppers (512×512)	7.5937	7.9991
Method 2 (triple image encryption)	Cameraman (256×256)	7.0097	7.9971
	Lena (256×256)	7.0115	7.9969
	Rice (256×256)	7.5690	7.9972

the two basic measures that are used to measure the key sensitivity in the algorithm. The key sensitivity is measured in two parts:

1. Key sensitivity in the encryption process
2. Key sensitivity in the decryption process

4.6.5.1 Key Sensitivity in the Encryption Process

Key sensitivity in the encryption process measures the sensitivity of keys in the encryption process. Figure 4.7 shows the key sensitivity results obtained from the "Lena" image by changing keys in Method 1 from x to $x+10^{-15}$, y to $y+10^{-15}$, a to $a+10^{-15}$, b to $b+10^{-15}$, c to $c+10^{-15}$, d to $d+10^{-15}$. In the similar way, other keys are also checked in Method 1. Table 4.7 shows the NPCR and UACI results obtained from the "Lena" image when the encryption keys are changed in Method 1. Figure 4.8 shows the key sensitivity results of the merged "Cameraman," "Rice," and "Lena" image obtained by changing keys in Method 2 from e to $e+10^{-15}$, f to $f+10^{-15}$. In a similar way, other keys are also checked in Method 2. Table 4.8 shows the NPCR and UACI results obtained from the

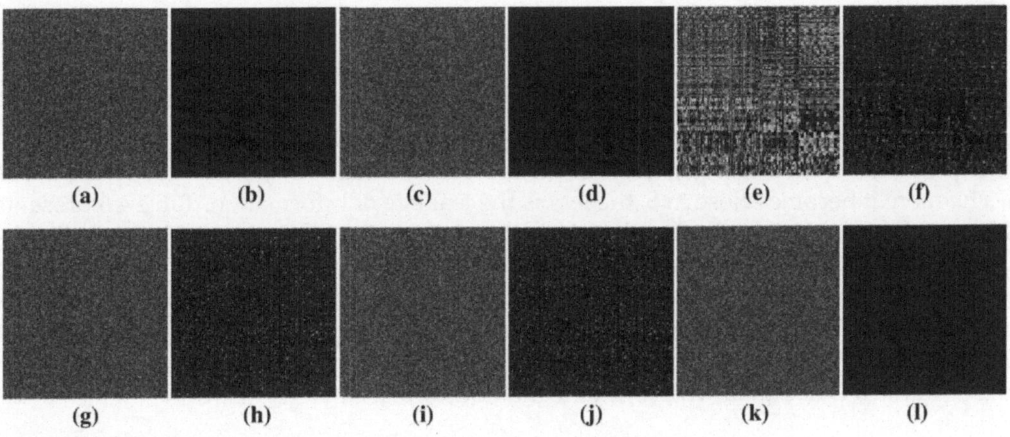

FIGURE 4.7
Key sensitivity results of "Lena" image using parallel sub-image encryption technique: encrypted images by changing key (a) x, (c) y, (e) a, (g) b, (i) c, (k) d; (b) difference image of (a) and Figure 4.3(b), (d) difference image of (c) and Figure 4.3(b), (f) difference image of (e) and Figure 4.3(b), (h) difference image of (g) and Figure 4.3(b), (j) difference image of (i) and Figure 4.3(b), (l) difference image of (k) and Figure 4.3(b).

TABLE 4.7

NPCR and UACI Results of "Lena" Image when the Encryption Keys are Changed in Method 1

Encryption keys	$x+10^{-15}$	$y+10^{-15}$	$a+10^{-15}$	$b+10^{-15}$	$c+10^{-15}$	$d+10^{-15}$
NPCR (%)	99.6017	99.6029	99.4196	99.4231	99.5182	99.5127
UACI (%)	33.7026	33.6261	31.2457	31.3334	32.4967	32.4880
Encryption keys	$p+10^{-15}$	$q+10^{-15}$	$r+10^{-15}$	$v+10^{-15}$	$E+10^{-15}$	$e+10^{-15}$
NPCR (%)	99.6097	99.6083	99.5127	99.5220	99.5898	99.6122
UACI (%)	33.7628	33.7422	31.6400	31.7765	32.4599	33.7765
Encryption keys	$ita+10^{-15}$	$u+10^{-15}$	$v+10^{-15}$	$m+10^{-15}$	$n+10^{-15}$	
NPCR (%)	99.6123	99.6024	99.6032	99.5890	99.5871	
UACI (%)	33.5343	33.4667	33.4567	32.9971	32.8961	

merged "Cameraman," "Rice," and "Lena" image when the encryption keys are changed in Method 2. The difference images of Figures 4.7 and 4.8 and the NPCR and UACI results in Tables 4.7 and 4.8 shows that more than 99% of pixels are changed by slightly changing the encryption keys. This concludes that in both the methods, the secret keys are highly sensitive to the algorithms.

4.6.5.2 Key Sensitivity in the Decryption Process

Key sensitivity in the decryption process measures the sensitivity of keys in the decryption process. Figure 4.9 shows the key sensitivity results obtained from the "Lena" image by changing keys in the Method 1 decryption process from x to $x+10^{-15}$, y to $y+10^{-15}$, a to $a+10^{-15}$, b to $b+10^{-15}$, c to $c+10^{-15}$, d to $d+10^{-15}$.

Table 4.9 shows the NPCR results obtained from the "Lena" image when the decryption keys are changed in Method 1. In a similar way, the other keys are also checked in

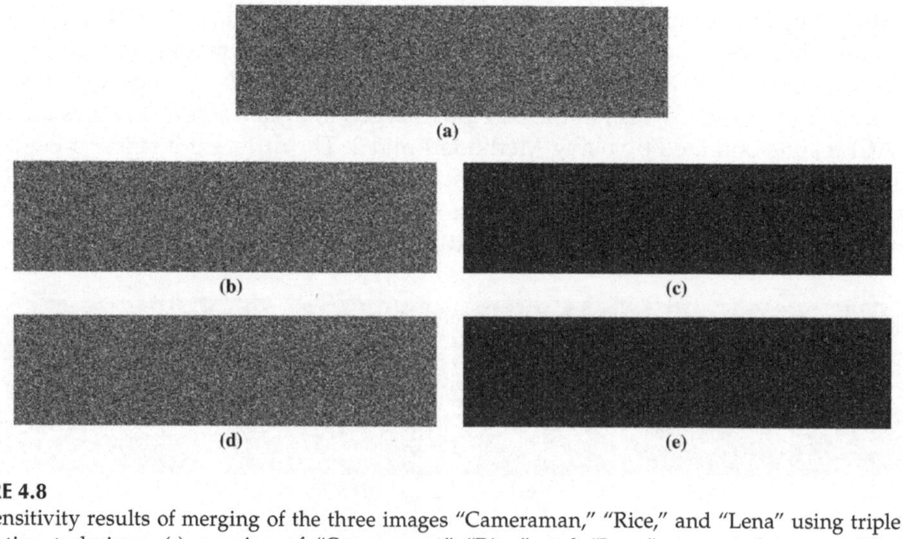

FIGURE 4.8
Key sensitivity results of merging of the three images "Cameraman," "Rice," and "Lena" using triple image encryption technique: (a) merging of "Cameraman," "Rice," and "Lena" encrypted images; (b) merged encrypted images by changing key e of Sine map, (c) difference image of (a) and (b), (d) merged encrypted images by changing key f of Tent map, (e) difference image of (a) and (d).

TABLE 4.8

NPCR and UACI Results of Merged "Cameraman," "Rice," and "Lena" Image when the Encryption Keys are Changed in Method 2

Encryption keys	$x+10^{-15}$	$r+10^{-15}$	$y+10^{-15}$	$s+10^{-15}$	$z+10^{-15}$	$t+10^{-15}$
NPCR (%)	99.6134	99.5882	99.6166	99.5864	99.6037	99.5826
UACI (%)	33.5727	32.4023	33.5899	32.3033	33.5243	32.4897
Encryption keys	$q+10^{-15}$	$u+10^{-15}$	$g+10^{-15}$	$ita+10^{-15}$	$e+10^{-15}$	$v+10^{-15}$
NPCR (%)	99.6196	99.5864	99.6197	99.5884	99.6276	99.6108
UACI (%)	33.5643	32.7654	33.5962	32.7701	33.6113	33.6443
Encryption keys	$f+10^{-15}$	$w+10^{-15}$				
NPCR (%)	99.6174	99.5994				
UACI (%)	33.5376	33.4118				

the Method 1 decryption process. Figure 4.10 shows the key sensitivity results obtained from the merged "Cameraman," "Rice," and "Lena" image by changing the keys in the Method 2 decryption process from e to $e+10^{-15}$, f to $f+10^{-15}$. In a similar way, other keys are also checked in the Method 2 decryption process. Table 4.10 shows the NPCR results obtained from the merged "Cameraman," "Rice," and "Lena" image when the decryption keys are changed in Method 2. Figures 4.9 and 4.10 shows that the encrypted images are not decrypted by small changes in the decryption keys. The NPCR results in Tables 4.9 and 4.10 show that more than 99% of pixels are changed by slightly changing the decryption keys. This shows that in both the methods, the secret keys are highly sensitive to the algorithms.

4.6.6 Plaintext Sensitivity Analysis

Plaintext sensitivity measures the sensitivity of pixels in the plaintext during the encryption process. Figure 4.11 shows the plaintext sensitivity results obtained from the "Lena" image for Method 1 by changing pixels values at positions (1, 1), (64, 64), (128, 128) and (256, 256) in the original image. Figure 4.12 shows the plaintext sensitivity results obtained from the merged "Cameraman," "Rice," and "Lena" image for Method 2 by changing pixels values at positions (20, 60) and (60, 320) in the original merged images. Table 4.11 shows the NPCR and UACI results obtained by using Methods 1 and 2. The different images in Figure 4.11 and Figure 4.12 and the NPCR and UACI results in Table 4.11 show that more than 99% of pixels are changed by slightly changing one pixel value in the plaintext. This concludes that in both the methods, the plaintexts are highly sensitive to the algorithms.

(a) (b) (c) (d) (e) (f)

FIGURE 4.9
Key sensitivity results of "Lena" image for Method 1 in the decryption process: decrypted images by changing key (a) x, (b) y, (c) a, (d) b, (e) c, (f) d.

FIGURE 4.10
Key sensitivity results of merged "Cameraman," "Rice," and "Lena" image for Method 2 in the decryption process: decrypted images by changing key (a) *e* in Sine map and (b) *f* in Tent map.

TABLE 4.9
NPCR and UACI Results of "Lena" Image when the Decryption Keys are Changed in Method 1 Decryption Process

Decryption keys	$x+10^{-15}$	$y+10^{-15}$	$a+10^{-15}$	$b+10^{-15}$	$c+10^{-15}$	$d+10^{-15}$
NPCR (%)	99.6123	99.6146	98.7624	98.8733	98.8346	98.9722
Decryption keys	$p+10^{-15}$	$q+10^{-15}$	$r+10^{-15}$	$v+10^{-15}$	$E+10^{-15}$	$e+10^{-15}$
NPCR (%)	99.6133	99.6170	98.6464	98.6799	98.9988	99.6109
Decryption keys	$ita+10^{-15}$	$u+10^{-15}$	$v+10^{-15}$	$m+10^{-15}$	$n+10^{-15}$	
NPCR (%)	99.6174	99.6128	99.6176	99.6089	99.6012	

TABLE 4.10
NPCR and UACI Results of Merged "Cameraman," "Rice," and "Lena" Image when the Decryption Keys are Changed in Method 2 Decryption Process

Decryption keys	$x+10^{-15}$	$r+10^{-15}$	$y+10^{-15}$	$s+10^{-15}$	$z+10^{-15}$	$t+10^{-15}$
NPCR (%)	99.6127	99.6057	99.6092	99.5967	99.6132	99.6091
Decryption keys	$q+10^{-15}$	$u+10^{-15}$	$g+10^{-15}$	$ita+10^{-15}$	$e+10^{-15}$	$v+10^{-15}$
NPCR (%)	99.6108	99.5990	99.6138	99.6077	99.6128	99.6077
Decryption keys	$f+10^{-15}$	$w+10^{-15}$				
NPCR (%)	99.6162	99.6006				

4.6.7 Known-Plaintext Attack and Chosen-Plaintext Attack Analysis

An image encryption scheme can be easily attacked by known-plaintext attack and chosen-plaintext attack if it is not very dependent on the original image. Both Methods 1 and 2 are highly dependent on the original images. This will be proved by considering the following points.

- First, all the secret keys used in Methods 1 and 2 are related to the original image through the SHA-256 hash algorithm.
- Second, the diffusion process is dependent on the PWLCM system, whose secret keys are dependent on the original images.

This shows that Methods 1 and 2 are highly image-dependent and capable of effectively repelling known-plaintext and chosen-plaintext attacks.

The chosen-plaintext attack can also be verified by observing the encrypted images of all white pixels and all black pixels in the chosen images. Figures 4.13 and 4.14 show the results of a chosen-plaintext attack for Methods 1 and 2, respectively. Both the figures

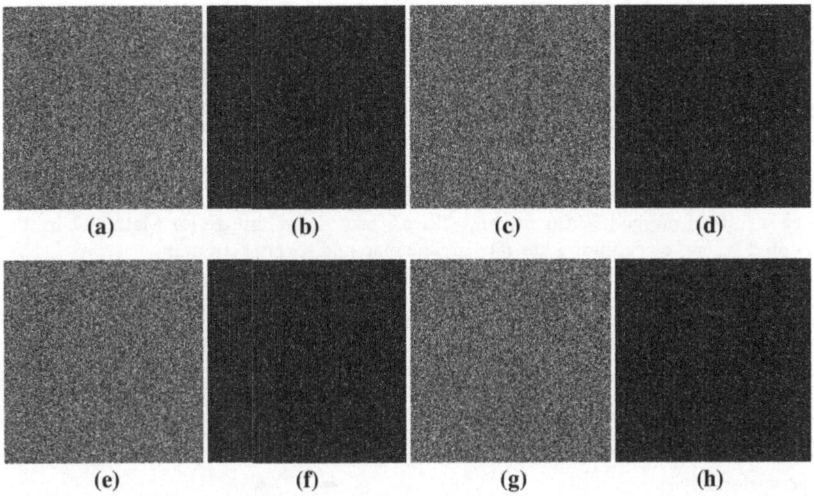

FIGURE 4.11
Plaintext sensitivity results of "Lena" image using Method 1: Encrypted images with changed pixel value at position (a)(1,1), (c)(64, 64), (e)(128, 128), and (g)(256, 256); (b) difference image of (a) and Figure 4.3(b), (d) difference image of (c) and Figure 4.3(b), (f) difference image of (e) and Figure 4.3(b), (h) difference image of (g) and Figure 4.3(b).

show that, by using any chosen images, the corresponding encrypted images cannot be identified.

4.6.8 Occlusion Attack Analysis

Many times, we come across insecure channels for transmission wherein our data gets tampered with in the process. This section analyses occluded data decryption. *Occluded* data means the data that get hidden or overridden in the transmission process [52]. We performed 1/16, 1/4 and 1/2 data occlusion of the cipher image using Method 1 and then analysed the decrypted image of it. Figure 4.15 shows the results of the occlusion attack. In Figure 4.15, it can be seen that the decrypted image of the 1/16 occlusion and the 1/4 occlusion is distorted but perceptible to human eye, but that the decrypted image of the 1/2 occlusion is not recoverable. So, this indicates that the encryption algorithm of Method 1

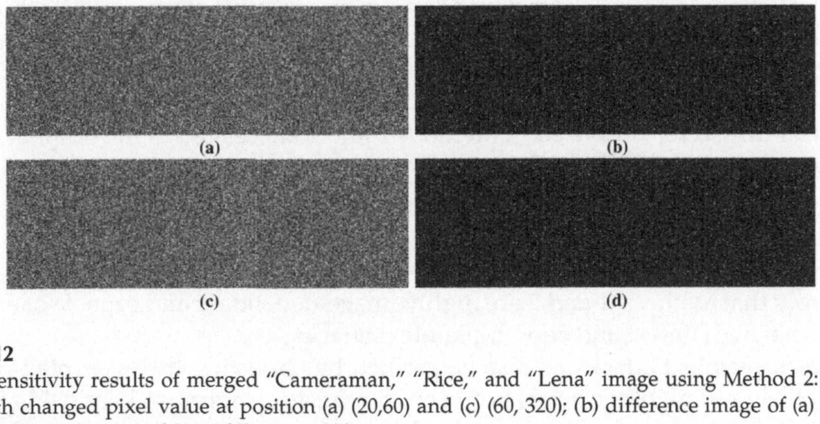

FIGURE 4.12
Plaintext sensitivity results of merged "Cameraman," "Rice," and "Lena" image using Method 2: Encrypted images with changed pixel value at position (a) (20,60) and (c) (60, 320); (b) difference image of (a) and Figure 4.8(a), (d) difference image of (c) and Figure 4.8(a).

TABLE 4.11

NPCR and UACI Results of "Lena" Image in Method 1 when the Pixels in Original Image are Changed and the NPCR and UACI Results of Merged "Cameraman," "Rice," and "Lena" Image in Method 2 when the Pixels in Merged Original Image are Changed

	Method 1				Method 2	
Pixel Positions	**(1,1)**	**(64,64)**	**(128,128)**	**(256,256)**	**(20,60)**	**(60,320)**
NPCR (%)	99.5973	99.6027	99.6078	99.6021	99.6013	99.6058
UACI (%)	33.4999	33.5121	33.4962	33.5018	33.5165	33.5128

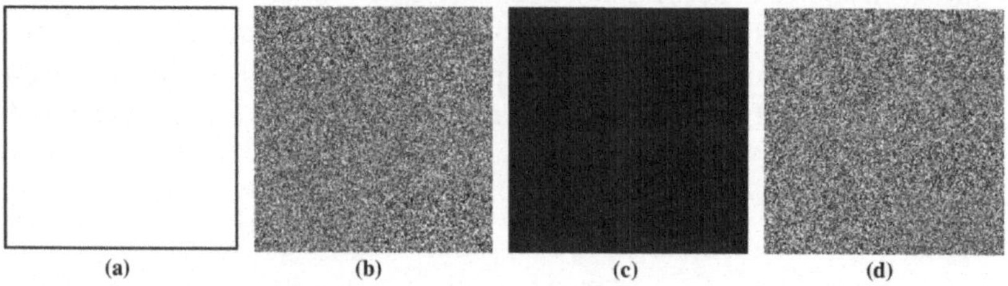

FIGURE 4.13
Chosen-plaintext attack analysis results of Method 1: (a) white image, (b) encrypted white image, (c) black image, (d) encrypted black image.

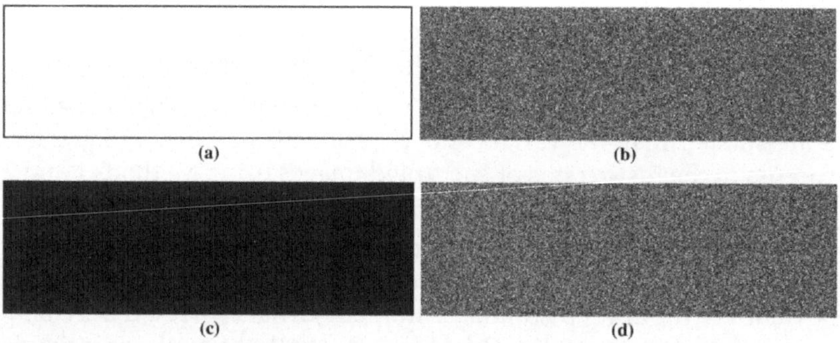

FIGURE 4.14
Chosen-plaintext attack analysis results of Method 2: (a) white image, (b) encrypted white image, (c) black image, (d) encrypted black image.

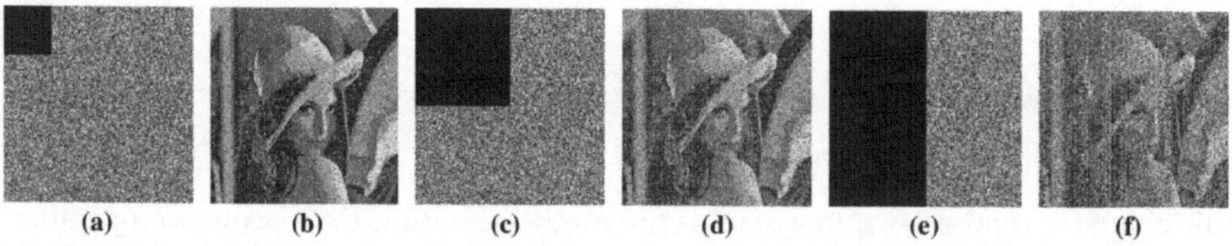

FIGURE 4.15
Occlusion attack analysis results of "Lena" image using Method 1: (a) encrypted image with 1/16 occlusion, (b) decrypted image of (a), (c) encrypted image with 1/4 occlusion, (d) decrypted image of (c), (e) encrypted image with 1/2 occlusion, (f) decrypted image of (e).

can withstand an attack of up to 25% occlusion. In a similar way, the occlusion attack can be checked for Method 2.

4.6.9 Time–Complexity Analysis

Both Methods 1 and 2 are simulated on a system with the following hardware configuration: 3.40GHz i7 processor, 4GB RAM, and Windows 10 operating system. MATLAB R2012a is used as a compiling software. The total time elapsed for executing the Method 1 encryption algorithm is 3.272–3.933 seconds for a 256×256 sized image and 5.273–5.564 seconds for a 512×512 sized image. This is a highly reduced execution time as DNA operations are performed on images. Similarly, the total time elapsed for executing the Method 2 encryption algorithm is 8.298–8.893 seconds for 256×256 sized images and 13.280–13.707 seconds for 512×512 sized images. This is also a highly reduced execution time, as DNA operations are performed on triple images. In both the methods, $\text{cell2mat}(x)$ and $\text{mat2cell}(y)$ functions take more operation time. Using general operations of converting cell to matrix and matrix to cell reduces the total encryption time. Hence, both the methods are time-efficient image encryption techniques.

4.6.10 Performance Comparison of Proposed Encryption Scheme with Existing Encryption Schemes

The comparisons of Methods 1 and 2 with other methods are as follows. In both the methods, the comparison is done on the "Lena" image.

- The total key spaces of both the methods are much larger than the key spaces of [47] which is $10^{42} \approx 1.4349 \times 2^{139}$ and [48] which is more than $10^{120} \approx 2^{400}$.
- The histogram variance results of the "Lena" encrypted image using Methods 1 and 2 are smaller than the variances of [47, 48]. This proves that a greater uniformity occurs in the histograms of encrypted images using Methods 1 and 2.
- The correlation coefficient values of the "Lena" encrypted image, including all the other images using Methods 1 and 2 and the other encryption schemes [14, 15], have nearer to zero along all the three directions. Moreover, the correlation coefficient values of the encrypted images in Method 1 and Method 2 are better than the encryption schemes [14, 15]. This shows that both the methods resist statistical attack more effectively in comparison to others.
- The NPCR and UACI values of the "Lena" image by using Methods 1 and 2 are larger than the expected values and also better than the encryption schemes [47, 48].

4.7 Conclusions and Future Works

This chapter proposes two image encryption schemes using DNA sequence operations and multiple chaotic maps. In the first scheme, a parallel sub-image operation-based image encryption is proposed. In the second scheme, a triple image encryption is proposed. Similar to first scheme, the second scheme is also a parallel sub-image operation-based image encryption, which is performed in each of the three images to speed up

the encryption process. Both the encryption schemes produce higher key space results to strongly resist brute-force attack. Apart from that, the encryption schemes resist all known common attacks, including known-plaintext attacks and chosen-plaintext attacks. The comparison results show that the proposed schemes perform better in comparison to the others. Finally, both the schemes take less time to simulate whole algorithms. In future, DNA and chaos-based image encryption techniques can be used for multiple images (more than three images). Cycle operation–based DNA sequence operations [49] and DNA insertion and deletion operations [50] can be used in the near future for multiple image encryptions.

Key Terminology and Definitions

Image security: This term refers to the protection or securing of images from attackers. Due to advancements in internet and digital technology, people are storing and transmitting larger numbers of digital images, and hence, image security is essential in the current digital era.

Parallel sub-image encryption: This term means "sub-image formation then parallel encryption"; that is, the whole image is divided into sub-images and then all the sub-images are processed separately and also differently. This process reduces the execution time of encryption and decryption. Moreover, this process is more confusing for the attacker.

Triple image encryption: Image encryption is a technique used to convert a meaningful (intelligible) image into a meaningless (unintelligible) image. Normal image encryption techniques are suitable for encrypting single images; however, for the encryption of multiple images together, single-image encryption techniques are not efficient. Hence, to overcome this problem, the use of multiple image encryption technique is advisable. Triple image encryption is a technique used to simultaneously encrypt three meaningful images into three meaningless images. This reduces the encryption time and hence increases the throughput.

DNA sequence operation: The DNA sequence consists of four nucleic acid bases: A (Adenine), T (Thymine), C (Cytosine), and G (Guanine). The operation on those four nucleic acid bases is called the *DNA sequence operation*. The three common DNA sequence operations that have been used to date in image encryption are the DNA encoding and decoding operation, DNA addition and subtraction operation, and DNA-XOR operation.

Chaotic maps: The term *chaos* means "a state of disorder." A chaotic map is a map that exhibits some sort of chaotic behaviour such as ergodicity, determinacy, non-periodicity, non-linearity, sensitivity to initial conditions and system parameters, etc. Chaotic maps are related to unpredictable behaviour of dynamic systems. The long-term prediction of chaotic systems is very difficult.

Security analysis: Security analysis analyses the security of an encryption algorithm. The commonly used security analyses used in image encryption are key space analysis, key sensitivity analysis, plaintext sensitivity analysis, statistical analysis, differential analysis, information entropy analysis, known-plaintext and chosen-plaintext attack analysis, occlusion attack analysis, timing complexity analysis, etc.

References

1. FIPS, P. (1977). Data Encryption Standard. Washington, DC: U.S. Department of Commerce, National Bureau of Standards.
2. FIPS, P. (1980). DES modes of operation: Category, ADP operations; subcategory, computer-security. Washington, DC: U.S. Department of Commerce, National Bureau of Standards.
3. Announcing the Advanced Encryption Standard (AES). (2001). Gaithersburg, MD: Computer Security Division, Information Technology Laboratory, National Institute of Standards and Technology.
4. Dang, P. and Chau, P. (2000). Image encryption for secure Internet multimedia applications. *IEEE Transactions on Consumer Electronics*, 46(3), 395–403. doi:10.1109/30.883383.
5. Zeghid, M., Machhout, M., Khriji, L., Baganne, A., and Tourki, R. (2007). A modified AES based algorithm for image encryption. *International Journal of Computer Science and Engineering*, 1(3), 70–75. doi:scholar.waset.org/1999.4/7580.
6. Gaj, K. and Chodowiec, P. (2001). Fast implementation and fair comparison of the final candidates for advanced encryption standard using field programmable gate arrays. *Topics in Cryptology – CT-RSA 2001 Lecture Notes in Computer Science*, 2020, 84–99. doi:10.1007/3-540-45353-9_8.
7. McLoone, M. and McCanny, J. V. (2003). Rijndael FPGA implementations utilising look-up tables. *Journal of VLSI Signal Processing Systems for Signal, Image and Video Technology*, 34(3), 261–275. doi:10.1023/A:1023252403567.
8. Kamali, S. H., Shakerian, R., Hedayati, M., and Rahmani, M. (2010). A new modified version of Advanced Encryption Standard based algorithm for image encryption. *2010 International Conference on Electronics and Information Engineering*, 1, V1–141. doi:10.1109/iceie.2010.5559902.
9. Gao, H., Zhang, Y., Liang, S., and Li, D. (2006). A new chaotic algorithm for image encryption. *Chaos, Solitons, and Fractals*, 29(2), 393–399. doi:10.1016/j.chaos.2005.08.110.
10. Samhita, P., Prasad, P., Patro, K. A. K., and Acharya, B. (2016). A secure chaos-based image encryption and decryption using crossover and mutation operator. *International Journal of Control Theory and Applications*, 9, 17–28.
11. Gupta, A., Thawait, R., Patro, K. A. K., and Acharya, B. (2016). A novel image encryption based on bit-shuffled improved tent map. *International Journal of Control Theory and Applications*, 9, 1–16.
12. Shadangi, V., Choudhary, S. K., Patro, K. A. K., and Acharya, B. (2017). Novel Arnold scrambling based CBC-AES image encryption. *International Journal of Control Theory and Applications*, 10, 93–105.
13. Matthews, R. (1989). On the derivation of a "chaotic" encryption algorithm. *Cryptologia*, 13(1), 29–42. doi:10.1080/0161-118991863745.
14. Liu, W., Sun, K., and Zhu, C. (2016). A fast image encryption algorithm based on chaotic map. *Optics and Lasers in Engineering*, 84, 26–36. doi:10.1016/j.optlaseng.2016.03.019.
15. Wang, X., Wang, S., Zhang, Y., and Guo, K. (2017). A novel image encryption algorithm based on chaotic shuffling method. *Information Security Journal: A Global Perspective*, 26(1), 7–16. doi:10.1080/19393555.2016.1272725
16. Fridrich, J. (1998). Symmetric ciphers based on two-dimensional chaotic maps. *International Journal of Bifurcation and Chaos*, 08(06), 1259–1284. doi:10.1142/s021812749800098x.
17. Wang, Y., Wong, K., Liao, X., and Chen, G. (2011). A new chaos-based fast image encryption algorithm. *Applied Soft Computing*, 11(1), 514–522. doi:10.1016/j.asoc.2009.12.011.
18. Zhang, L. Y., Hu, X., Liu, Y., Wong, K., and Gan, J. (2014). A chaotic image encryption scheme owning temp-value feedback. *Communications in Nonlinear Science and Numerical Simulation*, 19(10), 3653–3659. doi:10.1016/j.cnsns.2014.03.016.
19. Li, Y., Wang, C., and Chen, H. (2017). A hyper-chaos-based image encryption algorithm using pixel-level permutation and bit-level permutation. *Optics and Lasers in Engineering*, 90, 238–246. doi:10.1016/j.optlaseng.2016.10.020.

20. Zhang, Q., Guo, L., and Wei, X. (2013). A novel image fusion encryption algorithm based on DNA sequence operation and hyper-chaotic system. *Optik – International Journal for Light and Electron Optics*, 124(18), 3596–3600. doi:10.1016/j.ijleo.2012.11.018.
21. Zhang, Y., Wen, W., Su, M., and Li, M. (2014). Cryptanalyzing a novel image fusion encryption algorithm based on DNA sequence operation and hyper-chaotic system. *Optik – International Journal for Light and Electron Optics*, 125(4), 1562–1564. doi:10.1016/j.ijleo.2013.09.018.
22. Xie, T., Liu, Y., and Tang, J. (2014). Breaking a novel image fusion encryption algorithm based on DNA sequence operation and hyper-chaotic system. *Optik – International Journal for Light and Electron Optics*, 125(24), 7166–7169. doi:10.1016/j.ijleo.2014.07.111.
23. Guesmi, R., Farah, M. A., Kachouri, A., and Samet, M. (2015). A novel chaos-based image encryption using DNA sequence operation and Secure Hash Algorithm SHA-2. *Nonlinear Dynamics*, 83(3), 1123–1136. doi:10.1007/s11071-015-2392-7.
24. Enayatifar, R., Abdullah, A. H., Isnin, I. F., Altameem, A., and Lee, M. (2017). Image encryption using a synchronous permutation-diffusion technique. *Optics and Lasers in Engineering*, 90, 146–154. doi:10.1016/j.optlaseng.2016.10.006.
25. Wu, X., Kurths, J., and Kan, H. (2017). A robust and lossless DNA encryption scheme for color images. *Multimedia Tools and Applications*, 77(10), 12349–12376. doi:10.1007/s11042-017-4885-5.
26. Watson, J. D. and Crick, F. H. (1953). A structure for deoxyribose nucleic acid. *Nature*, 171(4356), 737–738.
27. Coelho, L. D. and Mariani, V. C. (2012). Firefly algorithm approach based on chaotic Tinkerbell map applied to multivariable PID controller tuning. *Computers and Mathematics with Applications*, 64(8), 2371–2382. doi:10.1016/j.camwa.2012.05.007.
28. Coelho, L. D. (2008). A quantum particle swarm optimizer with chaotic mutation operator. *Chaos, Solitons and Fractals*, 37(5), 1409–1418. doi:10.1016/j.chaos.2006.10.028.
29. Wang, L., Ye, Q., Xiao, Y., Zou, Y., and Zhang, B. (2008). An image encryption scheme based on cross chaotic map. *2008 Congress on Image and Signal Processing*, 3, 22–26. doi:10.1109/cisp.2008.129
30. Wang, X. and Yang, L. (2012). Design of pseudo-random bit generator based on chaotic maps. *International Journal of Modern Physics B* 26(32), 1250208. doi:10.1142/s0217979212502086.
31. Li, S., Chen, G., and Mou, X. (2005). On the dynamical degradation of digital piecewise linear chaotic maps. *International Journal of Bifurcation and Chaos*, 15(10), 3119–3151. doi:10.1142/s0218127405014052.
32. Mondal, B. and Mandal, T. (2017). A light weight secure image encryption scheme based on chaos & DNA computing. *Journal of King Saud University – Computer and Information Sciences*, 29(4), 499–504. doi:10.1016/j.jksuci.2016.02.003.
33. Zhou, Y., Bao, L., and Chen, C. P. (2014). A new 1D chaotic system for image encryption. *Signal Processing*, 97, 172–182. doi:10.1016/j.sigpro.2013.10.034.
34. Menezes, A. J., van Oorschot, P. C., and Vanstone, S. A. (2001). *Handbook of Applied Cryptography*. Boca Raton, MA: CRC Press. eBook ISBN: 9781439821916.
35. Patidar, V., Pareek, N., Purohit, G., and Sud, K. (2011). A robust and secure chaotic standard map based pseudorandom permutation-substitution scheme for image encryption. *Optics Communications*, 284(19), 4331–4339. doi:10.1016/j.optcom.2011.05.028.
36. Floating-point Working Group. IEEE Standard for Binary Floating-Point Arithmetic. ANSI. IEEE Std. 1985, 754–1985.
37. Shannon, C. E. (1949). Communication theory of secrecy systems. *Bell System Technical Journal*, 28(4), 656–715. doi:10.1002/j.1538-7305.1949.tb00928.x.
38. Brindha, M. and Gounden, N. A. (2016). A chaos based image encryption and lossless compression algorithm using hash table and Chinese Remainder Theorem. *Applied Soft Computing*, 40, 379–390. doi:10.1016/j.asoc.2015.09.055.
39. Chai, X., Chen, Y., and Broyde, L. (2017). A novel chaos-based image encryption algorithm using DNA sequence operations. *Optics and Lasers in Engineering*, 88, 197–213. doi:10.1016/j.optlaseng.2016.08.009.

40. Zhang, J., Fang, D., and Ren, H. (2014). Image encryption algorithm based on DNA encoding and chaotic maps. *Mathematical Problems in Engineering*, 2014, 1–10. doi:10.1155/2014/917147.
41. Enayatifar, R., Abdullah, A. H., and Isnin, I. F. (2014). Chaos-based image encryption using a hybrid genetic algorithm and a DNA sequence. *Optics and Lasers in Engineering*, 56, 83–93. doi:10.1016/j.optlaseng.2013.12.003.
42. Zhu, C. (2012). A novel image encryption scheme based on improved hyperchaotic sequences. *Optics Communications*, 285(1), 29–37. doi:10.1016/j.optcom.2011.08.079.
43. Wang, X., Zhang, Y., and Bao, X. (2015). A novel chaotic image encryption scheme using DNA sequence operations. *Optics and Lasers in Engineering*, 73, 53–61. doi:10.1016/j.optlaseng.2015.03.022.
44. Shannon, C. E. (2009). A mathematical theory of communication, Edition: 1, Wiley-IEEE Press, pp. 968. ISBN: 9780470544242.
45. Wei, X., Guo, L., Zhang, Q., Zhang, J., and Lian, S. (2012). A novel color image encryption algorithm based on DNA sequence operation and hyper-chaotic system. *Journal of Systems and Software*, 85(2), 290–299. doi:10.1016/j.jss.2011.08.017.
46. Yi, X., Tan, C. H., and Siew, C. K. (2002). A new block cipher based on chaotic tent maps. *IEEE Transactions on Circuits and Systems I: Fundamental Theory and Applications*, 49(12), 1826–1829. doi:10.1109/tcsi.2002.805714.
47. Zhu, Z., Zhang, W., Wong, K., and Yu, H. (2011). A chaos-based symmetric image encryption scheme using a bit-level permutation. *Information Sciences*, 181(6), 1171–1186. doi:10.1016/j.ins.2010.11.009.
48. Zhang, Y. and Wang, X. (2014). A symmetric image encryption algorithm based on mixed linear–nonlinear coupled map lattice. *Information Sciences*, 273, 329–351. doi:10.1016/j.ins.2014.02.156.
49. Hu, T., Liu, Y., Gong, L. H., and Ouyang, C. J. (2017). An image encryption scheme combining chaos with cycle operation for DNA sequences. *Nonlinear Dynamics*, 87(1), 51–66. doi:10.1007/s11071-016-3024-6.
50. Hu, T., Liu, Y., Gong, L. H., Guo, S. F., and Yuan, H. M. (2017). Chaotic image cryptosystem using DNA deletion and DNA insertion. *Signal Processing*, 134, 234–243. doi:10.1016/j.sigpro.2016.12.008.
51. Yin, Z., Pan, L., Fang, X. (Eds.). (2013). Proceedings of the Eighth International Conference on Bio-Inspired Computing: Theories and Applications (BIC-TA), 212. eBook ISBN: 978-3-642-37502-6. doi:10.1007/978-3-642-37502-6.
52. Patro, K. A. K., and Acharya, B. (2018). Secure multi-level permutation operation based multiple colour image encryption. *Journal of Information Security and Applications*, 40, 111–133. doi:10.1016/j.jisa.2018.03.006.

5

Application of DNA Computing in the Cloud Computing Environment

Debashree Devi

CONTENTS

5.1 Introduction

The evolution of computers from main-frame computers to quantum computers involves up-grading large-scale machines to affordable and less expensive computers through the use of integrated circuits as the processing module (Harris, 2010; Huth and Chebula, 2011). Cloud computing is a benchmark in the era of large-scale computation that offers computation-as-a-service, which minimises the need to purchase a server, storage space and network hardware (Huth and Chebula, 2011). Cloud computing defines a computing archetype that provides sharing of resources, servers, storage space and hardware devices

over a network through the assimilation of a number of systems (Harris, 2010). The term *cloud* signifies its capability of providing cloud services remotely.

Security in the cloud environment is one of the challenges that is causing most concern. Generally in cloud computing, users are not concerned about the service sources or about the storage of data, as cloud services are provided without any information about the underlying architecture. This acts as a loophole that raises many issues about security in the cloud environment. In (Zhou et al., 2010), a number of examples are given about security breaches that have occurred in the cloud environment, some of which are stated below:

1. In March 2009, Google Docs found a flaw that inadvertently shares users' documents.
2. In October 2007, a Salesforce.com employee fell victim to a phishing attack and leaked a customer list, which generated further targeted phishing attacks.
3. Epic.com lodged a formal complaint to the Federal Trade Commission (FTC) against Google for its privacy practices in March 2009. EPIC was successful in an action against Microsoft Passport.

Apart from the security issue, the secure storage of large volumes of users' sensitive data is another challenge in cloud environment. The trend of high-scale computation leads to the generation of larger volumes of data, which demands advanced management and processing tools. Data confidentiality and data integrity are two prime factors in data storage in a cloud environment (Sukumaran and Mohammed, 2018). Various encryption algorithms are implemented to provide data confidentiality, while digital signature techniques are used to ensure the integrity of the stored data. Recently, a number of research projects have focused on this aspect (Namasudra, 2017; Namasudra and Roy, 2015, 2016, 2017a,b,c, 2018; Namasudra et al., 2014, 2017a,b,c; Sarkar et al., 2015). The DNA computing technique can also be used to define the confidentiality of users' sensitive data being stored in the cloud server, as it enables data storage in the form of DNA bases. The encryption and storage of data in the form of DNA bases can solve both the issues of data security and secure data storage.

One of the prominent solutions for cloud data security is *cryptography*. The main goal of cryptography is to transmit the data or message (file) between the sender and receiver over an *untrusted* medium in such a manner that an attacker or malicious user is unable to read the original data content (Kuo, 2018). Based on the way the encryption and decryption keys are used, a cryptosystem can be divided into two types:

1. *Symmetric key encryption*: In this type of cryptosystem, the encryption and decryption processes are executed by using the same key (Kumar and Wollinger, 2006; Schneier, 1996), and the key has to be shared between the sender and receiver.
2. *Asymmetric key encryption*: Here, different keys or values are used to encrypt and decrypt a data or message, and this method does not require sharing keys between the sender and receiver. Each party is provided with a pair of keys which contains (1) a private key, to be used in encryption; and (2) a public key, to be used for decryption.

Technically, asymmetric key encryption is more secure than symmetric key encryption.

Steganography is another approach to data security, and it provides data-hiding techniques. The technique of steganography involves an algorithm that embeds the data or

information into a carrier, and develops an identifier function that returns the embedded data or information (Channalli and Jhadav, 2009). The function uses a secret key, and this is only shared with the authorised user. The function recovers and identifies the existence of the data or information within the carrier by using the secret key. The definition of the secret key is a prime aspect of this technique.

DNA computing is a computing area wherein DNA, molecular biology, hardware and biochemistry are used to encode genetic details in computers. DNA was first proposed in the field of computation by Leonard Adleman in 1994 at the University of Southern California. Adleman has demonstrated a proof-of-concept by using DNA in the form of a computation that solved the seven-point "Hamiltonian Path Problem" (Adleman, 1994). Initially, this novel approach was used to solve NP-hard problems. However, very soon it was found to be ineffective as a solution to NP-hard problems. Computer scientist Mitsunori Ogihara and biologist Animesh Ray described an implementation of Boolean circuits by using DNA computing in 1999 (Ogihara and Ray, 1999). Afterwards, a number of techniques were developed by using DNA computing to solve different problems, such as the SAT problem, 0-1 planning problem, integer planning problem, graph theory, optimisation problem, database, cryptography, etc. Currently, DNA computing is one of the trending fields in biology and computer science. DNA computing provides a tool for parallel computing, and it takes advantage of different molecules of DNA (Lewin, 2002). The complexity and organisation of all human beings is mainly based on the coding system of the four components of the DNA molecule. One strand of DNA consists of four bases (A, T, C and G). After attaching with the deoxyribose, those nucleotides can act together as a string for generating long sequences (Boneh and Lipton, 1996). Different DNA structures are used by researchers for solving different problems (Lipton, 1995). There are many operations in DNA computing models, namely *adjoining*, *cut*, *insertion*, *paste* and *deletion*. The benefits of DNA computing are:

1. Speed
2. Minimal storage requirements
3. Minimal power requirements

Moving on to security concerns, recently *DNA cryptography* has become an emerging trend, developed based on DNA computing (Tornea and Borda, 2009). Here, DNA computing is used for data encryption and data storage for achieving a strong security mechanism, so that unauthorised, malicious users and attackers are unable to retrieve the data content. Many cryptographic algorithms have been proposed based on DNA, namely symmetric and asymmetric key cryptosystems using DNA, triple stage DNA cryptography, DNA steganography systems, DNA-based chaotic computing and DNA-based encryption algorithms. In terms of the pros of DNA cryptography, it is much more secure than conventional cryptography methods, as it is quite difficult to pre-guess the exact ordering of DNA bases. Besides, a large volume of data can be stored in the DNA bases, which reduces the cost of large storage space.

The main contribution of this chapter is to explore the applications of DNA computing in a cloud computing environment for solving issues of cloud data security. In this chapter, a DNA cryptography model is proposed for providing secure accessing of data from the cloud server and secure and scalable storage of cloud data. A user-attribute–based DNA key generation technique is proposed for generating the secret key. A DNA-based data encryption algorithm is defined for generating the ciphertext. Performance and security

analyses of the proposed work are carried out in comparison with three existing DNA cryptography models.

5.2 Related Work

A number of cryptographic methods have been developed for secure transmission of data. Shamir (1985) first proposed Identity Based Encryption (IBE). Here, an identity is specified by the sender of any data, and this must be matched by the receiver for data decryption. A fuzzy identity–based encryption model was proposed after few years, wherein the identity of the user is presented with a set of descriptive attributes. A respective user can only decrypt the encrypted data if the attributes are exactly matched with the ciphertext's attributes. Attribute Based Encryption (ABE) was proposed by Sahai and Waters (Sahai and Waters, 2005) for providing complex data accessing. Ciphertext Policy Based ABE (CPABE) and Key Policy Based ABE (KPABE) are two main types of ABE. In CPABE, the ciphertext is assigned to an access policy, and private keys for the authorised users are generated on the basis of users' attributes (Bethencourt et al., 2007). An authorised user can decrypt the ciphertext if the attributes of the user satisfy the assigned access policy in the ciphertext. In KPABE, the ciphertext is assigned with attributes, and an access structure is associated with the user's private key, which denotes the user's identity (Goyal et al., 2006). Here, an authorised user can decrypt the ciphertext if his/her access tree satisfies all the respective attributes of the ciphertext. Most of the traditional methods are based mainly on tough complex mathematical equations, where researchers have mainly focused on increasing the complexity by modifying the equations (Lin and Hsueh, 2008; Liu and Tsai 2007; Sencar et al., 2007). In some traditional approaches, to improve the security, the data is usually encrypted and the encrypted data embedded into various multimedia data or files as the cover media, like audio, video, image, etc. (Voloshynovskiy et al., 2003; Barni et al., 2003). Thus, malicious users or attackers are unable to find the data to break the cryptosystems.

In recent years, many researchers have implemented DNA cryptography to improve the security of data. (Abbasy et al., 2011) proposed a method for data hiding using DNA sequences. They proposed this scheme for increasing complexity by using the view of software. (Khalifa and Atito, 2012) proposed a steganography scheme for a secure data exchanging process. This scheme supports transforming a plaintext into a collection of amino acids, and then encrypting it with a DNA-based playfair cipher. Hsu and Lee (2006) used three encryption processes to improve data security, namely the insertion method, the complementary method and the substitution method. In the insertion method, the basic technique is modified by introducing a random number generator that divides the DNA sequence into a number of segments. Complementary pair strings of different lengths are the core concept that is adapted in the second method. In the substitution method, the DNA sequence is transformed into an encrypted DNA sequence by checking the identical binary bits between the DNA sequence and the original message. All these three techniques enhance the secure transmission of data. Gehani et al., (2000) presented the fundamentals of DNA-based cryptography schemes. They have analysed the one-time-pad concept in comparison to the DNA steganography scheme. Implementation of one-time pad allows the encryption process to be quite unpredictable for malicious users or hackers. Gehani et al. also described DNA-based cryptography in terms of two techniques: the substitution approach with random paired mapping and the XOR scheme

with nucleotide computation. The encrypted DNA sequence is appended with secret keys to be transmitted among other DNA strands. (Wang and Zhang, 2009) proposed a novel technique for using DNA computing with the RSA algorithm in the field of information security. They combined the DNA computing theory with asymmetric key cryptography. There are five phases in this scheme:

1. Setup plaintext
2. Matching with nucleotides
3. Converting nucleotides to numbers
4. Encrypting message
5. Recovering plaintext

(Zhang and Gao, 2015) proposed a novel technique for data hiding that provides encoding for secure information and cover media, and an algorithm is used to generate the codon table, which controls the substitution. (Gupta and Singh, 2015) proposed a DNA-based encryption scheme. There are two main processes in this scheme, namely the substitution process and the Central Dogma of Molecular Biology (CDMB). In the substitution process, a DNA sequence is randomly taken from 163 million (approximately) those are publicly available in a DNA database. In CDMB, the DNA sequence is transformed into a Ribonucleic Acid (RNA) sequence, and then, the RNA sequence is transformed into protein sequence. (Cui et al., 2008) proposed a DNA-based encryption system by using DNA synthesis, PCR amplification and DNA digital encoding with conventional cryptography methods. The primers and coding scheme are used as the key, while digital encoding and standard cryptography are used for data encryption. (Ahmed and Ibrahim, 2017) proposed a new data encryption technique, DNA-DES, by integrating DNA cryptography with standard Data Encryption System (DES) to increase the key space and the number of possible permutations of the DES method.

5.3 Background Studies

Biologically, DNA (or Deoxyribonucleic Acid) is a long molecule that contains the genetic information of our body (DNA, 2016). DNA is responsible for everything about the cell in a human body, from its birth to growth and then to its demise. The characteristics of one's DNA are inherited from one's parents. DNA is made of nucleotides, and each of those has three units: a phosphate group, a sugar molecule called *deoxyribose* and a nitrogen base. There are four nitrogen bases, namely Adenine (A), Cytosine (C), Guanine (G) and Thymine (T). The pairing of nitrogen bases is fixed, i.e., Adenine (A) is always paired with Thymine (T), and Cytosine (C) is paired with Guanine (G) (DNA, 2016; Amos, 2008). The random combinations of DNA bases facilitate the generation of a variety of DNA sequences. The NCBI database provides a huge dataset for DNA sequences (Wheeler *et al.*, 2004). In cryptography, the codons, primers and pairing concepts of DNA are employed in different ways to improve data security.

In DNA, *codons* are a triplet of adjacent nucleotides that corresponds to a certain amino acid. A *primer* is a DNA sequence of generally ten sequence pairs. It represents the starting point of DNA synthesis. A primer is needed for DNA replication. Primers are added

to the DNA sequence, which are termed *keys*. The concept of a primer is incorporated into the DNA-based key generation to enhance the complexity and randomness of the data encryption process.

5.4 System Model and Requirements

5.4.1 System Model

The system model for the proposed scheme consists of three entities, namely the cloud service provider, the data owner and the users.

1. *Cloud service provider (CSP):* The CSP can be defined as the backbone of the proposed architecture. The CSP is responsible for providing cloud services to both users and data owner (DO) and facilitates access to various cloud services, providing a number of servers equipped with storage space and computational capability. The CSP is responsible for keeping track of various information such as the size of the data, the latest access details for each particular data, etc. The CSP generates the certificates and system parameters required for a successful transaction.
2. *Data owner (DO):* The DO can be any organisation or user who wants to store data on the cloud server. For maintaining the data properly on the cloud server, the DO depends on the CSP.
3. *Users:* Users can be defined as entities who can access cloud services from the cloud server. The authenticity of the users is verified before they are permitted to access the cloud services.

5.4.2 System Requirements

The requirements of the proposed model are explained below:

1. *Fine grained access control*: Fine grained access control allows for the deployment of a security policy to manage users' accountability in the cloud server without hindering the activities of various vendors. The CSP ensures that only authorised users can have the access rights or policies for the data they are requesting. The DO provides the access rights or polices to the authorised users for accessing the cloud services. For example, VMware has recently launched a product named vShield Zones Product, which enforces a firewall-protected network and port connections on the virtual server, with a transparent association of security policies and firewall rules in the virtual environment (Strom, 2010).
2. *Security*: The security of users' sensitive data is a must for a cloud system. Sharing of data over the network opens the trapdoor for various malicious users who are hovering over the internet, waiting for the chance to gain unauthorised access to the cloud server. The CSP must ensure the secured storage and accessing of users' sensitive data by defining some security policies and authentication rules.
3. *Reduced system overhead*: The cloud services are computationally complex, which leads to an increase in the system overhead. Employing strong security polices in the

cloud server requires the deployment of data encryption or data hiding algorithms with complex intermediate stages, making it difficult enough to be hacked by the hackers or malicious users. Implementation of these techniques in the cloud server increases the system overhead. It is essential to reduce the number of intermediate operations in the security techniques in order to reduce the system overhead.

Based on the system requirements, the design goals for the proposed scheme can be defined as:

1. To define a strong data security model, providing secure storage for and access to users' sensitive data in the cloud server
2. To reduce the intermediate stages within the security model, so that computational complexity can be minimised, and thus a reduced system overhead achieved
3. To define a more scalable storage architecture in the cloud server

5.5 Proposed Scheme

There are two key phases in the proposed scheme, namely:

1. DNA-based data storage phase.
2. DNA-based data encryption stage, which provides secure storing and sharing of data in the cloud server. Data are stored in the CSP in terms of DNA bases, which will facilitate the storage of huge volumes of data within a small storage space.

A DNA-based data encryption scheme is employed with a user attribute–based key generation phase, which facilitates the secure sharing of users' sensitive data in the cloud server. The system architecture for the proposed scheme is presented in Figure 5.1.

In the proposed scheme, there are five phases:

1. System setup
2. User registration
3. User authorisation
4. DNA-based data storage
5. Data access phase

The mechanism involved in each phase is discussed in the following sections.

5.5.1 System Setup phase

In this phase, the CSP is the driving force that initiates the execution of a user's request, setting up the necessary credentials, which include private and public key pair for the CSP, and private and public key pairs for the DO. The key pairs are distributed to the users at the time of user registration. The distribution of keys is performed in such a manner that the user's public key is accessible to all, while the public key of the DO is accessible only to

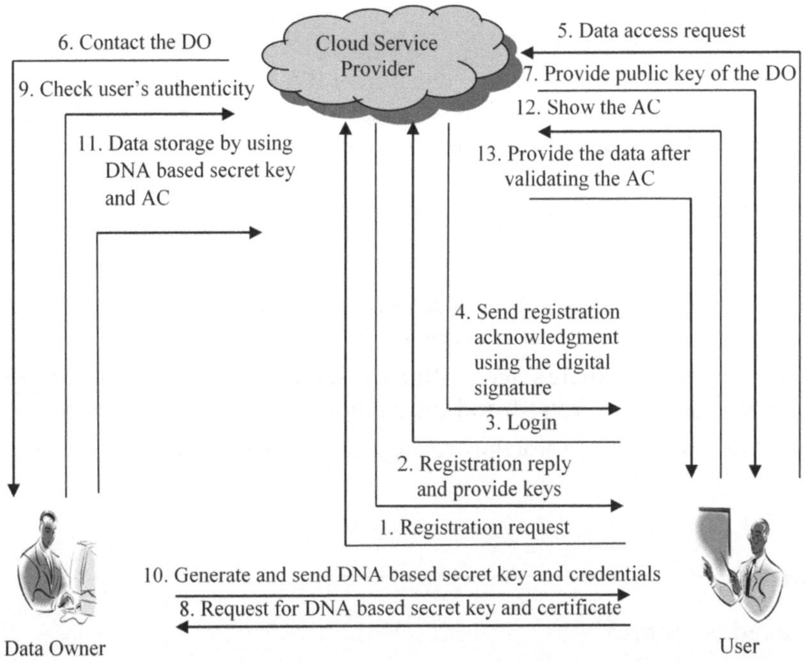

FIGURE 5.1
System architecture of the proposed model.

the authorised users. The public key of the CSP is accessible to the DO and the authorised users. The private keys are kept confidential. The concept of the pseudo-random number generator is to select a large prime number, p to define the multiplicative group, Z_p^*. The uniqueness of keys is the prime aspect of security of the proposed scheme.

5.5.2 User Registration Phase

This phase is put into action whenever the cloud server receives a registration request from a user. The CSP gathers all the information about the user to make a profile of that user. A duplicate identity is generated for establishing a secure communication gateway by using Secure Socket Layer (SSL), and a package is delivered to the user with the duplicate identity, registration acknowledgement and public–private key pairs. The package is transmitted to the respective registered/authorised user, who retrieves the credentials by means of the generated duplicate identity and his/her real identity.

5.5.3 User Authorisation Phase

In this phase, the registered users are given permission to login into the system after requesting data access permission. The CSP, on receiving the user's request, sends an acknowledgement to the user, encrypted with the digital signature. The user, on receiving the acknowledgment, becomes eligible to request data access. The access request received by the CSP is forwarded to DO, and the users are provided with the public key of the DO. Users can communicate with the DO through the public key received and requests for the DNA-based key sequence, i.e., the secret key of the DO and the Access Certificate (AC). The authorisation of the user is verified by the DO by retrieving the user's profile from

the CSP. Based on the user's attributes, a 256-bit encryption key is generated by the DO, which is provided to the user along with the AC. The user, on receiving the DNA-based key sequence, becomes fully authorised to access data in the cloud server.

5.5.4 DNA-Based Data Storage Phase

The process of data storage in the cloud server is executed by the DO in conjunction with the CSP. The DO can store data in the cloud server once it is registered at the CSP. The DO registration and authorisation phases are identical to those of the user. In the proposed architecture, the concept of DNA computing is employed for achieving secure data storing in the cloud server. The process of data storage can be divided into two sub-phases:

1. DNA-based key generation and key distribution
2. DNA-based data encryption.

The details of each sub-phase are discussed below.

5.5.4.1 DNA-Based Key Generation and Key Distribution

In this phase, the DO performs the DNA-based key generation and distribution based on the user's attributes. The process is initiated on receiving the user's request for the secret key and the AC, and it is executed only for the authorised users, once their authenticity has been verified by the DO. The DO retrieves the user's profile from the CSP to select the attributes for DNA-based key generation. A set of operations is performed on the user-attribute values to convert them into equivalent DNA bases. The various operations are described below:

1. *Decimal encoding*: Decimal encoding allows the conversion of the user-attribute values into equivalent decimal format. Table 5.1 presents the values for decimal encoding.
2. *Decimal-to–binary conversion*: Through this operation, the decimal numbers obtained are converted to their 8-bit binary equivalent. The binary string is scaled to meet the length of defined key (256 bits), either by appending zeros at the right of the string or by deleting the extra bits.
3. *Complementary rule*: The binary string obtained is converted to its complementary binary string.
4. *Binary DNA encoding*: Binary DNA encoding allows the conversion of the complementary binary string to DNA bases, based on the values given in Table 5.2.
5. *DNA complementary rule*: The achieved DNA sequence is converted to its complementary DNA bases by using DNA complementary rules, given as: A⟶. G, G⟶. C, C⟶. T, and T⟶. A

The result of these operations is obtained as a DNA sequence that holds the user-attribute values in the form of DNA bases. The next step involves calculating the primer. To calculate the primer, a random DNA base is selected from the standard UCBI/NCBI database, which comprises a massive collection of 160 million DNA bases. Now, say the length of the selected DNA base is L, and it is divided into two equal parts, each with a length of L/2.

TABLE 5.1

Decimal Encoding Table

Alphabets/ Symbols	Decimal Value	Alphabets/ Digits	Decimal Value	Alphabets/ Digits	Decimal Value
A	01	K	11	U	21
B	02	L	12	V	22
C	03	M	13	X	23
D	04	N	14	Y	24
E	05	O	15	Z	25
F	06	P	16	,	26
G	07	Q	17	-	27
H	08	R	18	:	28
I	09	S	19	.	29
J	10	T	20	[] (blank space)	30

TABLE 5.2

Binary DNA Encoding Table

Binary String	DNA Bases
00	A
01	T
10	C
11	G

These two parts are then added to the extreme left and right side of the user-attribute value based resultant DNA sequence. The resultant DNA sequence after primer addition is then converted to its binary equivalent, according to the binary DNA encoding rules (Table 5.2). The DO selects a random number by using a pseudo-random number generator, which is converted into equivalent binary format, and the length is then adjusted according to the length of the resultant binary DNA sequence. The DO performs the EXOR operation with the resultant binary string and the binary DNA sequence. The result of the EXOR operation yields the final generated key.

For key distribution, the CSP sends the generated key and the AC, along with other essential credentials, i.e., complementary rule, primer, binary DNA encoding rule, and decimal encoding rule, to the user. The key, AC and the other credentials are encrypted by the DO by using the DO's private key and the user's public key prior to transmission to the user. At the user end, the user can retrieve the key, AC and the other credentials by decrypting the package by using the user's private key and the DO's public key. An illustrative example is presented below to visualise the DNA-based key generation phase.

Step 1: The user sends a request to the DO for accessing data in the cloud server.

Step 2: The DO checks the user's authenticity. If the user is an authorised user, then they proceed to the next step. Otherwise, the user's request is terminated.

Step 3: Once the user's authenticity is confirmed, the DO fetches the respective user's attributes from the CSP. Say, the DO fetches the following attribute values:

USR_ID = U05

DATE OF BIRTH = 03-11-1988

Step 4: The attribute values are put together in a single sequence, as

Sub-phase 1: U0503111988

Sub-phase 2: 2105…… [By using decimal encoding]

Sub-phase 3: 0001010100000011…………00001000 [8-bit binary equivalent of each decimal value]

Sub-phase 4: 11101010111111………0111 [By using binary complementary rule]

Sub-phase 5: GCC……TG [By using binary DNA encoding]

Sub-phase 6: CTT…AC [By using DNA complementary rules]

Sub-phase 7: Select a random DNA base, say 01000111010001

Sub-phase 8: Divide the DNA base into two equal parts and add it to both sides of the result of sub-phase 6.

$$\underbrace{01000110 \mid 10100011}_{01000110|CTT\ldots AC|10100011}$$

Sub-phase 9: 01000110|100101…0010|10100011. (By using reverse binary DNA encoding)

Sub-phase 10: 0001010100000011…00001000 ⊕ 01000110|100101……0010|10100011.

Sub-phase 11: 100100….

Sub-phase 12: CTA…. (Final key)

5.5.4.2 DNA-Based Data Encryption

The basic mechanism employed in this phase is the EXOR operation between the plaintext and the DNA key sequence. For performing EXOR, the DO first converts the plaintext into an 8-bit binary format. The resultant binary string is then converted into a long binary string of 256 bits by appending zeros at the right. The generated DNA key sequence is also converted into a 256-bit-long binary string. The 256-bit-long binary plaintext and the DNA key sequence is divided into four blocks each with 64 bits. The EXOR operation is performed for each 64-bit block of plaintext and DNA key sequence for four cycles. After the first cycle, the DNA key sequence is rotated by using the DNA complementary rule and left-key shifting to get the DNA key sequence for the next cycle. In the next cycle, EXOR is performed between the rotated DNA key sequence and the result of the first cycle. This process continues for up to four cycles. For each execution of the EXOR operation, a new DNA key sequence is generated, which makes the data encryption more secure. For each 8-bit combination of the resultant binary string, randomly created ASCII (American Standard Code for Information Interchange) values are assigned, as given in Table 5.3. This introduces randomness in the encryption process. Each ASCII value is then converted into the equivalent binary format. Binary DNA encoding (Table 5.2) is employed on the resultant binary string to obtain the DNA sequence. The generated DNA sequence is altered by using the DNA complementary rule, to generate the ciphertext (CT).

The generated CT is then encrypted by the DO's private key and the CSP's public key and is transmitted to the CSP along with the Access Certificate (AC). For each data in the cloud server, a distinct AC is assigned by the CSP, which makes it more secure. The CSP decrypts the message by using the DO's public key and the CSP's private key and retrieves the CT. The CT is then stored in the cloud server and is provided to the user on receiving a request for data accessing.

TABLE 5.3
Randomly Created ASCII Table Related to Each 8-Bit Binary Value

8-Bit Binary Value	ASCII
00000000	81
00000001	251
00000010	87
00000011	119
.	.
.	.
.	.
11111111	7

5.5.5 Data Access Phase

In the data access phase, authorised users are able to access the data in the cloud server by using the DNA-based secret key and AC. In the proposed architecture, once an user is authorised, he/she is always granted as authorised user. The CSP will not ask the user to obtain the secret key and the AC every time they request for data accessing. The secret key and the AC are provided to the user when he/she requests to access data for the very first time. The user has the privilege of accessing data on the basis of this information in future. The data accessing phase involves the following steps:

1. The DO initiates a data processing task once it receives a data accessing request from the user. As the initial step, the DO sends the encrypted public key of the DO to the user, encrypted by using the DO's private key and the user's public key. At the user end, the user retrieves the DO's public key by decrypting the message.
2. In the next step, the user sends an encrypted message to the DO, requesting the DNA-based secret key and the AC for data accessing. The encryption involves user's private key and the DO's public key. At the DO's end, the message is decrypted by using the DO's private key and the user's public key. On getting the user request, the DO verifies the authenticity of the user and retrieves the user's profile from the CSP. Based on the attributes of the user obtained from the profile, the DO generates a 256-bit DNA-based key sequence, as depicted in Section 5.5.4.1. The DO encrypts the key, the AC and other credentials and transmits them to the user as an encrypted message. The public–private key pair used is same as the ones used in the previous step.
3. The user retrieves the DNA-based secret key and the AC by decrypting DO's message with user's private key and the DO's public key. To retrieve the original 256-bit key, the user has to perform all the operations evolved in the DNA-based key generation in a reverse fashion. The obtained AC is then forwarded to the CSP, encrypted by using the user's private key and the CSP's public key. At the CSP end, the CSP retrieves the AC by decrypting the message, and its legitimacy is checked with respect to the requested data. If the legitimacy of the AC is proved, the CSP performs a search for the data in the cloud server or in the database. If the AC is found to be invalid, the request is declined.

4. The data obtained from the cloud server are then encrypted by using DNA-based data encryption, as discussed in Section 5.5.4.2., and a CT is generated. The generated CT is transmitted to the user, encrypted with the CSP's private key and the user's public key.
5. The encrypted CT is decrypted at the user end by using the CSP's public key and the user's private key. The user now having the DNA-based key sequence, and other credentials, is capable of retrieving the original text/ message by performing DNA-based data decryption.

5.6 Performance Analysis

5.6.1 Cloud Simulation Environment

For performance evaluation, a cloud simulation environment needs to be set up. A popular tool available for cloud simulation is CloudSim (Calheiros et al., 2011), which was developed by a group of researchers at Melbourne University. The basic architecture of CloudSim has four layers of processing modules, namely cloud services, cloud resources, user interface structures and virtual machine services. The CloudSim toolkit is also equipped with a number of entities, such as Cloud Information Services (CIS), host, data centre, Virtual Machine (VM), cloudlet, broker and Virtual Machine Manager (VMM).

The experimental environment set up by using the CloudSim toolkit for implementing the proposed scheme and the existing schemes has the following features:

1. *Hardware requirement:*

   A desktop with 3.40 GHz Intel corei7 processor, 64 GB RAM and 4 TB hard disk.
2. *Software requirement:*
   a. Windows 10 operating system
   b. Apache Tomcat 9.0.5
   c. JAVA version 9
   d. CloudSim 3.0.3

For the simulation, the CloudSim toolkit is configured with 2000 physical node, of HP ProLiant ML 110 G4 and G5 servers. Each of the nodes are allocated with 32GB storage space and a network bandwidth of 2GB/sec. The HP ProLiant ML family of servers feature flexible, expandable tower storage, which sets an ideal architecture for remote sharing or data centres, providing a large amount of internal storage.

There are few modifications in the CloudSim for the simulation process. To achieve the dynamic resource distribution, a few classes are modified, namely *VMScheduler, VM, powerhost* and *CloudletScheduler.* A new class has been added in the CloudSim, namely *dynamicmemory,* which helps to guess the essential resources. After the cloudlet execution, the *dynamicmemory* class updates the resource distribution of Virtual Machines (VMs).

5.6.2 Results and Discussions

The basic objectives of the proposed architecture are to generate a strong security model, with secure accessing and storage of user-sensitive data; and to obtain scalable storage space.

In accordance with the objectives, a number of experiments can be done to evaluate the proficiency of the proposed architecture. For experimentation purposes, three existing algorithms are considered:

1. A DNA technology–based encryption scheme (Cui et al., 2008)
2. A data hiding method based on DNA coding and modulo-N operations (Zhang and Gao, 2015)
3. A data encryption standard with DNA cryptography (Ahmed and Ibrahim, 2017)

The following acronyms are considered for representing the existing schemes during simulation Table 5.4:

Different scenarios are considered with different parameters for the evaluation task. The proficiency of the proposed scheme is evaluated with the aspects of key retrieval time of user, data encryption time, data decryption time and average central processing unit (CPU) utilisation of the models.

Figure 5.2 depicts the performance of the models for key retrieval time with a varying number of users. The process of key retrieval in the proposed architecture involves a series

TABLE 5.4
Acronyms for Existing Methods

Algorithm	Acronyms
A DNA technology–based encryption scheme	DNAEn
Data hiding method based on DNA coding and modulo-N operations	DHDNAModN
Data encryption standard with DNA cryptography	DNADES

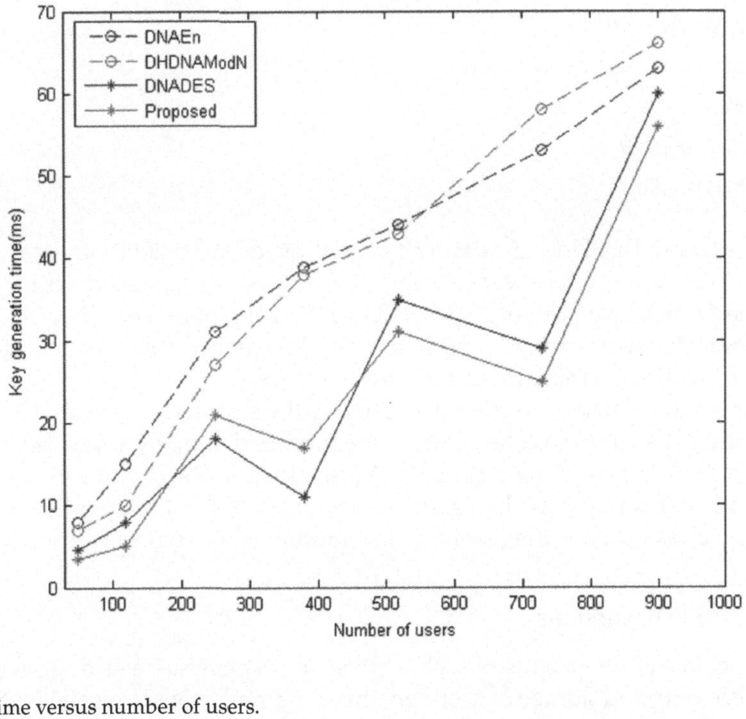

FIGURE 5.2
Key retrieval time versus number of users.

of operations to be carried out when the key is generated for the first time for a user request. It yields an increasing in key retrieval time at the initial, but decreased in later, as key retrieval time is same for old authorised users. In a cloud server, one can't predetermine the type of incoming user. Sometimes, the incoming user might be an old user and sometimes it might be a new user. So, the proposed scheme gives a zigzag curve for key retrieval time. In case of the DNAEn and DHDNAModN methods, conventional methods of key retrieval are followed, which need to be performed for each user, for each execution. Thus, it will record a linearly increasing curve for the DNAEn and DHDNAModN techniques. In case of DNADES, a random key is selected which might vary from user to user. However, the same key can be used by a specific user for accessing data in subsequent requests. Thus, the performance of DNADES for key retrieval time is recorded in a zigzag fashion.

The models are then evaluated for data encryption time with respect to an increasing number of users. In case of DNAEn, DHDNAModN and DNADES, the process of data encryption involves many intermediate stages. For every user request, the data has to be encrypted by going through all the intermediate stages. In case of the proposed scheme, the data encryption phase takes the DNA key sequence and the plaintext as input and performs EXOR operation followed by left-key shifting. In the proposed scheme, the generation of CT requires less stages than those in existing schemes. Besides, once a DNA key sequence is obtained for a specific user, it does not have to be fetched again in future. After user authorisation, the plaintext can be directly encrypted by using the DNA key sequence. Hence, the data encryption time of the proposed scheme is less than existing schemes. Figure 5.3 presents the experimental results for the data encryption time with respect to increasing number of users.

Next, the models are evaluated for data decryption time with respect to varying number of users. In case of the existing DNAEn, DHDNAModN and DNADES methods, the phase of data decryption involves the same number of intermediate stages as the data encryption

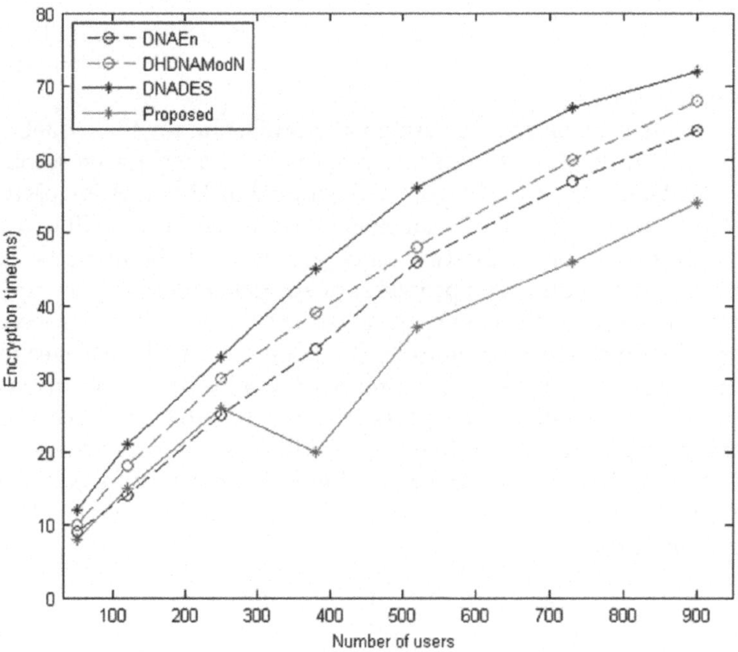

FIGURE 5.3
Data encryption time versus number of users.

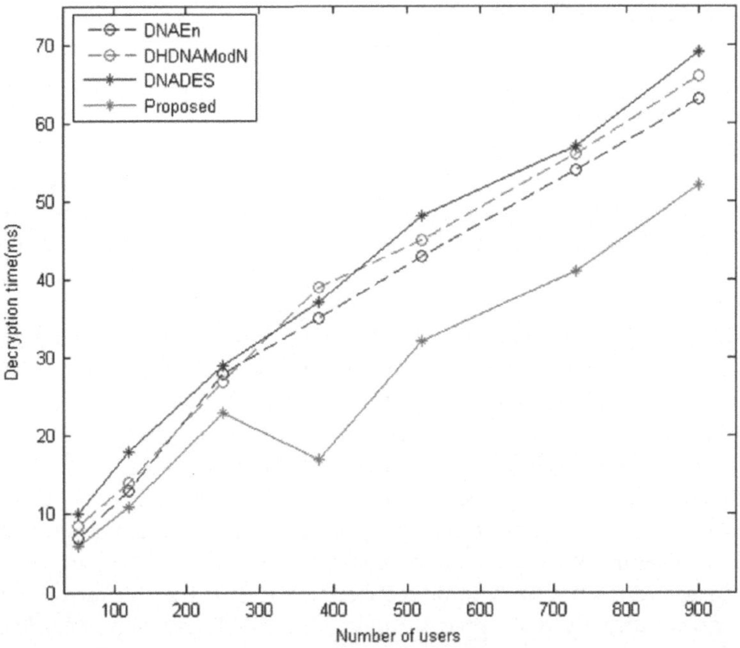

FIGURE 5.4
Data decryption time versus number of users.

phase. The proposed model also employs the same stages as DNA-based data encryption. Besides, the proposed scheme has the privilege of using a DNA key sequence that does not need to be retrieved for data decryption as it is already available to the user. Hence, the proposed scheme shows better performance than existing schemes with less data decryption time. The experimental results of data decryption time for an increasing number of users are presented in Figure 5.4.

The next experiment involves calculating CPU utilisation for the models. For calculating the rate of CPU utilisation, each experiment is performed for 60 seconds. The CPU utilisation is calculated for every 5 seconds, i.e., a total of 12 values are obtained for each experiment. The average of these 12 values is taken as measure of CPU utilisation. The computational complexity of the existing schemes is found to be more, as the underlying mechanisms are quite complex. The processes of key generation and key retrieval have to be executed for every user in the case of the existing schemes. The phases of data encryption and decryption in existing schemes evolve many intermediate stages for generating and retrieving the CT. In the case of the proposed scheme, the user attribute–based key generation and DNA-based data encryption facilitate the provision of secure data accessing by generating the secret key only once. It reduces the computational complexity of the proposed scheme as compared to the existing DNAEn, DHDNAModN and DNADES techniques. Figure 5.5 presents the experimental results for the rate of CPU utilisation for varying number of users.

5.6.3 Performance Analysis

1. *Fine grained access control*: In the proposed scheme, the DO employs a user attribute–based encryption technique to generate the secret key. The information

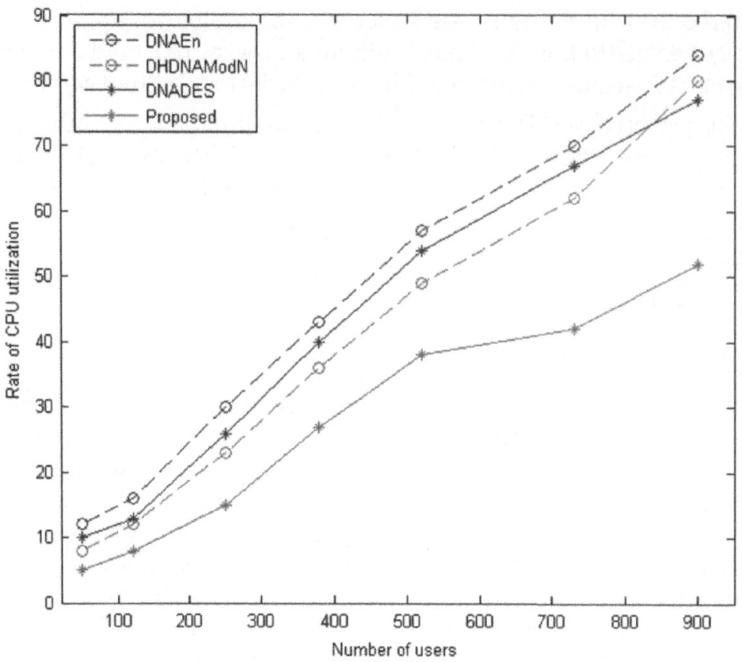

FIGURE 5.5
Rate of CPU utilisation versus number of users.

about the user's attributes is stored by the CSP. The DO fetches the user's profile at the time of key generation only, and the DO does not have any prior knowledge about the user. The same scenario is maintained at the user end. The user gets the DNA-based secret key and the AC from the DO with the consent of CSP. No prior information about the DO is available to the user. This strategy used in the proposed scheme establishes the double-blindness property of cloud computing.

2. *Secure data accessing*: There are two scenarios of data accessing. The first is when a new user requests permission of data access. For a new user, the DNA-based secret key is generated, based on the user's attributes, by the DO. For data accessing, a new user must be registered at the CSP. Once a user is proved to be authorised, the DO provides the DNA-based secret key and the AC. The second case is when an old user sends a request for data accessing. For an old user, the CSP directly provides the secret key and the AC after user authorisation. This user attribute–based key generation facilitates secure accessing of cloud data/services with a minimal computational overhead.
3. *Security*: In the proposed scheme, a user attribute–based DNA key generation technique is employed. The information about the user attributes is stored at the CSP at the time of user registration. Hence, only the CSP has exact knowledge about the user attributes. It is almost impossible for any malicious user to guess the exact values of the user attributes. Thus, the proposed scheme is secure against malicious users in the cloud server. In the data encryption phase, the DNA -based key sequence is used to generate the CT by using the EXOR operation and left-key shifting. The ordering of DNA bases for the CT is randomly defined by the DO; thus, it is difficult to guess the correct ordering of the DNA bases. Moreover, the other essential credentials used in the key generation and data encryption phases

are shared only with the authorised user. It is impossible for hackers or malicious users to get access to the cloud data without all these credentials. Hence, the proposed system is secure against any kind of attacks in the cloud server.

4. *Reduced system overhead*: The DOs are not required to be constantly online in the proposed scheme. They can come online to provide the AC and the DNA-based secret key to the users and can go offline after providing the credentials. Therefore, the system overhead is reduced in the proposed scheme.

5.6.4 Security Analysis

5.6.4.1 Man-in-the Middle Attack

The proposed scheme can provide security against man-in-the middle attacks. In the proposed scheme, when a user makes a request for the secret key and access certificate, the DO first checks the user's authenticity through the CSP. The necessary credentials are provided to authorised users only. If the authenticity of the user is not proved, the DO has the right to decline the user request. Moreover, the secret key and the AC is encrypted with the DO's private key and user's public key before being transmitted to the user. In the case of a man-in-middle attack, even if a hacker retrieves the message in between the transmission, he/she is unable to decrypt it without the private–public key pair. The same procedure is carried out in the transmission of CT to the CSP. An intruder can't retrieve the original message until he/she knows the private–public key pair of the DO and CSP.

5.6.4.2 Password Guessing Attack

In the proposed scheme, the DOs use the DNA-based secret key to encrypt a large volume of data, which is generated based on the user's attributes or any secret information of the user. Since an attacker or malicious user is unable to guess the secret information of an authorised user or all the attributes, the attacker cannot guess the password. Moreover, the DO uses binary complementary rules, DNA complementary rules, DNA encoding, DNA binary encoding and random ASCII value assignment for data encryption. These credentials are shared only with the authorised user after checking their authenticity at the CSP's end. Therefore, an attacker or hacker must guess all the attributes of an authorised user and s/he must have the secret keys, the AC and all other credentials for decrypting the data. So, the proposed scheme can be considered secure against a password guessing attack.

5.6.4.3 Internal Attack

An internal attack may be carried out by an employee or user or even by the CSP. It may also be done by the third party that supports the operations of the cloud service. In the proposed scheme, the DO generates a 256-bit DNA-based secret key for authorised users only. The generated key is very secure as it is generated based on the user-attribute values, which are stored only at the CSP. The DO has no knowledge about the user information. On the other hand, the details of other essential credentials used during the key generation and data encryption are unknown and are not shared with the CSP by the DO. All these processes introduce a high degree of randomness into the whole key generation and data encryption processes, thus making them more secure against internal attacks.

5.7 Conclusion

This chapter provides a detailed overview of various security issues in the cloud computing environment. The application of DNA computing for improving data security is studied in detail. A number of DNA-based data encryption schemes are discussed from the perspective of cloud data security. A DNA-based data security model is proposed in this chapter to facilitate DNA-based secure accessing of cloud data with scalable data storage. Though experimental analysis, the proposed architecture is proven to be effective in reducing computational overhead of the cloud system. The security analysis of the proposed scheme establishes its proficiency in repelling most of the attacks/threats encountered in a cloud server. Through this chapter, a comprehensive investigation of the effectiveness of DNA computing is presented, which can be used to build/define more effective data security models in future.

References

Abbasy, M. R., Manaf, A. A., and Shahidan, M. A. (2011). Data hiding method based on DNA basic characteristics. In E. Ariwa and E. E. Qawasmeh, (Eds.), *Digital Enterprise and Information Systems*, Springer: Berlin Heidelberg, pp. 53–62.

Adleman, L. M. (1994). Molecular computation of solutions to combinatorial problems. *American Association of Advancement of Science*, 266(5187), 1021–1024.

Ahmed, K. and Ibrahim, E. H. (2017). Increasing robustness of data encryption standard by integrating DNA cryptography. *International Journal of Computers and Applications*, 39(2) 91–105.

Amos, M. (2008). *DNA Computing. Encyclopedia of Complexity and System Science*, 4, 2089–2014.

Barni, M., Bartolini, F., and Furon, T., (2003). A general framework for robust watermarking security. *Signal Processing*, 83(10), 2069–2084.

Bethencourt, J., Sahai, A., and Waters, B. (2007). Ciphertext-policy attribute based encryption. In *Proceedings of the 2007 IEEE Symposium on Security and Privacy*, pp. 321–334.

Boneh, D. and Lipton, R. (1996). Making DNA computers error resistant. In *Proceedings of Second Annual Conference on DNA Based Computers*, Princeton, New Jersey, pp. 102–110.

Calheiros, R. N., Ranjan R., and Beloglazov A. (2011). Retrieved 05th January, 2018, from CloudSim: A toolkit for modeling and simulation of cloud computing environments and evaluation of resource provisioning algorithms. *Software-Practice and Experience (SPE)*, 41(1), 23–50.

Channalli, S. and Jhadav, A. (2009). Steganography: An art of hiding data. *International Journal on Computer Science and Engineering*, 1(3) 137–141.

Cui, G., Quin, L., Wang, Y., and Zhang, X. (2008). An encryption scheme using DNA technology. *BIC-TA*, 2008, 37–42.

DNA. (2016). Retrieved 20th January, 2018, from https://simple.wikipedia.org/wiki/DNA.

Gehani, A., LaBean, T., and Reif, J. (2000). DNA-based cryptography. In N. Jonoska, G. Paun, and G. Rozenberg, (Eds.), *Aspects of Molecular Computing*, Springer: Berlin Heidelberg, pp. 167–188.

Goyal, V., Pandey, O., Sahai, A., and Waters, B. (2006). Attribute-based encryption for fine-grained access control of encrypted data. In *Proceedings of the 13th Conference on Computer and Communications Security*, pp. 89–98.

Gupta, R. and Singh, R. K. (2015). An improved substitution method for data encryption using DNA sequence and CDMB, In *Proceedings of the 3rd International Symposium, SSCC 2015*, Kochi, India, pp. 197–06.

Harris, T. (2010). Cloud computing – an overview. Retrieved 12th January, 2018, from http://www.thbs.com/downloads/Cloud-Computing-Overview.pdf

Hsu, H. Z. and Lee, R. C. T. (2006). DNA based encryption methods. In *Proceedings of the 23rd Workshop on Combinatorial Mathematics and Computation Theory*, NantouHsies, Taiwan, pp. 145–150.

Huth, A. and Chebula, J. (2011). *The Basics of Cloud Computing*. Carnegie Mellon University.

Khalifa, A. and Atito, A. (2012). High-capacity DNA-based steganography. In *Proceedings of the 8th International Conference on Informatics and Systems (INFOS2012)*, Cairo, Egypt, pp. 76–80.

Kumar, S. and Wollinger, T. (2006). Embedded security in cars. In K. Lemke, C. Paar, and M. Wolf (Ed.), *Fundamentals of Symmetric Cryptography*. Springer: Berlin Heidelberg, pp. 125–143.

Kuo, C.-J. *Cryptography*. National Taiwan University: Taipei, Taiwan. Retrieved 15th February, 2018, from http://disp.ee.ntu.edu.tw/meeting/%E6%94%BF%E9%8C%A6/Cryptography/Cryptography.pdf

Lewin, D. I. (2002). DNA computing. *Computing in Science Engineering*, 4(3), 5–8. DOI:10.1109/5992.998634.

Lin, C. C. and Hsueh, N. L. (2008). A lossless data hiding scheme based on three-pixel block differences. *Pattern Recognition*, 41(4), 1415–25.

Lipton, R. J. (1995). DNA solution of hard computational problems. *Science*, 268(5210), 542–545.

Liu, T. Y. and Tsai, W. H. (2007). A new steganographic method for data hiding in Microsoft Word documents by a change tracking technique. *IEEE Transactions on Information Forensics and Security*, 2(1), 24–30.

Namasudra, S. (2017). An improved attribute-based encryption technique towards the data security in cloud computing. *Concurrency and Computation: Practice and Exercise*. DOI:10.1002/cpe.4364

Namasudra, S. and Roy, P. (2015). Size based access control model in cloud computing. In *Proceedings of the International Conference on Electrical, Electronics, Signals, Communication and Optimization*, IEEE, Visakhapatnam, India, pp. 1–4.

Namasudra, S. and Roy, P. (2016). Secure and efficient data access control in cloud computing environment: A survey. *Multiagent and Grid Systems-An International Journal*, 12(2) 69–90.

Namasudra, S. and Roy, P. (2017a). Time saving protocol for data accessing in cloud computing. *IET Communications*, 11(10), 1558–1565.

Namasudra, S. and Roy, P. (2017b). A new secure authentication scheme for cloud computing environment. *Concurrency and Computation: Practice and Exercise*, 29(20). DOI:10.1002/cpe.3864

Namasudra, S. and Roy, P. (2017c). A new table based protocol for data accessing in cloud computing. *Journal of Information Science and Engineering*, 33(3), 585–609.

Namasudra, S. and Roy, P. (2018). PpBAC: Popularity based access control model for cloud computing. *Journal of Organizational and End User Computing*, 30(4), 14–31.

Namasudra, S., Nath, S., and Majumder, A. (2014). Profile based access control model in cloud computing environment. In *Proceedings of the International Conference on Green Computing, Communication and Electrical Engineering*, IEEE, Coimbatore, India, pp. 1–5.

Namasudra, S., Roy, P., and Balamurugan, B. (2017a). Cloud computing: Fundamentals and research issues. In *Proceedings of the 2nd International Conference on Recent Trends and Challenges in Computational Models*, IEEE, Tindivanam, India.

Namasudra, S., Roy, P., Balamurugan, B., and Vijayakumar, P. (2017b). Data accessing based on the popularity value for cloud computing. In *Proceedings of the International Conference on Innovations in Information, Embedded and Communications Systems (ICIIECS)*, IEEE, Coimbatore, India.

Namasudra, S., Roy, P., Vijayakumar, P., Audithan, S., and Balamurugan, B. (2017c). Time efficient secure DNA based access control model for cloud computing environment. *Future Generation Computer Systems*, 73, 90–105.

Ogihara, M. and Ray, A. (1999). Simulating Boolean circuits on a DNA computer. *Algorithmica*, 25, 239–250.

Sahai, A. and Waters, B. (2005). Fuzzy identity-based encryption. In R. Cramer (Ed.), *Advances in Cryptology*, Springer, pp. 457–473.

Sarkar, S., Saha, K., Namasudra, S., and Roy, P. (2015). An efficient and time saving web service based android application. *SSRG International Journal of Computer Science and Engineering (SSRG-IJCSE)*, 2(8), 18–21.

Schneier, B. (1996). *Applied Cryptography*, 2nd edn., Wiley.

Sencar, H.T., Ramkumar, M., Akansu, A.N., and Sukerkar A. (2007). Improved utilization of embedding distortion in scalar quantization based data hiding techniques. *Signal Processing* , 87(5), 877–890.

Shamir, A. (1985). Identity-based cryptosystems and signature schemes. In G. R. Blakley and D. Chaum (Eds.), *Advances in Cryptology*, Springer, pp. 47–53.

Strom, D. (2010). How secure is the Cloud? Retrieved 10th February, 2018, from http://www.tomshardware.com/reviews/cloud-computing-security,2829-3.html

Sukumaran, S. C. and Mohammed, M. (2018). DNA cryptography for secure data storage in cloud. *International Journal of Network Security*, 20(3) 447–454.

Tornea, O. and Borda, M. E. (2009). DNA cryptographic algorithms. MEDITECH 2009, *IFMBE Proceedings* 26, pp. 223–226.

Voloshynovskiy, S., Pun, T., Fridrich J., Pérez-González, F., and Memon, N. (2003). Security of data hiding technologies. *Signal Processing*, 83(10), 2065–2067.

Wang, X. and Zhang Q. (2009). DNA computing-based cryptography. In *Proceedings of the 4th IEEE International Conference on Bio-Inspired Computing*, Beijing, China, pp. 1–3.

Wheeler, D. L., Church, D. M., Edgar, R., Federhen, S., Helmberg, W., Madden, T. L., Pontius, J. U., Schuler, G. U., Schriml, L. M., Sequeira E., Suzek T. O., Tatusova T. A., and Wagner L. (2004). Database resources of the National Center for Biotechnology Information: Update. *Nucleic Acids Research*, 32, 35–40. DOI:10.1093/nar/gkh073.

Zhang, S. and Gao, T. (2015). A novel data hiding scheme based on DNA coding and module-N operation, *International Journal of Multimedia and Ubiquitous Engineering* 10(4), 337–344.

Zhou, M., Zhang, R., Xie, W., Qian, W., and Zhou, A. (2010). Security and privacy in cloud computing: A survey. *Sixth International Conference on Semantics , Knowledge and Grids*, 105–112.

6

Taxonomy of Security Attacks in DNA Computing

Selvaraj Karthikeyan, Patan Rizwan, and Balusamy Balamurugan

CONTENTS

6.1 Introduction

Data security attacks by intruders are becoming widespread wherever there is a chance to insert malicious codes into a system or network. Whenever a computer is connected to the internet, it is at the risk of threat from viruses, worms and attacks from hackers. In the information technology (IT) sector, data security is the process of ensuring that the information stored in any device cannot be accessed without proper authorization. Data encryption changes meaningful information (UbaidurRahman et al., 2015) into new undefined information that is traceable only when an individual knows the decrypting key.

Cyberspace is becoming a dangerous place for all organizations and individuals who want to protect their sensitive data or reputation, as *n* number of people and machines can access it.

DNA is deoxyribonucleic acid (Saghir and Megherbi, 2015), used for storing genetic details in a cellular organism. There are four different bases: Adenine (A), Guanine (G),

Cytosine (C), and Thymine (T), where adenine must combine with Thymine and Cytosine must combine with Guanine and vice versa; these combinations are called "Watson Crick Complementarity" (Cui et al., 2009). DNA is made up of biological macromolecules constructed by nucleotides that have a single base, T.

DNA computing is the process of carrying out computations using biological molecules instead of silicon chips. It is an emerging area that converts the data in a particular problem to DNA oligonucleotides. Biomolecule tools are used for the decryption of DNA that is encrypted to obtain a computation result.

The first DNA computing was done by Leonard Adleman (Adleman, 1995) in 1994, when he solved seven nodes of the Hamiltonian Path problem by utilizing oligonucleotides. The objective was to make use of chemistry to figure out the problems that traditional computers cannot solve. DNA computing uses DNA characteristics for doing parallel computation (Cui et al., 2007). The research has since spread into various fields like biology, chemistry, and mathematics.

A computation is simply the execution of an algorithm, which is the step-by-step execution of instructions that takes the input, processes it, and gives an output. This is the sequence DNA computing will perform for storing any information without using the binary alphabet (1 and 0) used by ordinary computers.

It goes vastly beyond the storage capacities of ordinary electronic and optical media. A gram of DNA contains almost 10^{21} DNA bases or 10^{8} terabytes, so a few grams of DNA have the ability to store all the data in the world (Cui et al., 2007).

The three basic steps that are used in performing DNA computing are

1. Convert all the solutions to computational problems by encoding
2. Produce all the data that represent a possible solution to the computational problem
3. Pull out the problem solved with the support of Polymerase Chain Reaction (PCR)

DNA computing theories represent one way to apply cryptography (Ning, 2009). The art of cryptography security is to ensure that third parties cannot read messages by encoding them. The RSA algorithm (Wang and Zhang, 2009) that belongs to asymmetric key cryptography can be used for the encryption process. DNA cryptography is a grouping of cryptographic methods that have emerged with the mechanism of DNA computing, where DNA is the carrier of the source of information and the biological technology can be used as an implementation tool (UbaidurRahman et al., 2015).

DNA is used in cryptography in addition to computation, and it also stores and transfers data (UbaidurRahman et al., 2015; Cui et al., 2009). DNA computing is efficient for storing the data, but to store the huge volume of data, it needs high-technology laboratories. Hence, DNA cryptography security is tough to implement (UbaidurRahman et al., 2015). Algorithms such as DES, RSA, and MD5 are used to secure the data in *n* number of ways, but an intruder will be still able to break the algorithm and steal the data. To overcome the shortcomings of existing cryptography algorithms, DNA computing is used with cryptography and steganography (Gehani et al., 2003) to make an unbreakable algorithm.

A *security* mechanism is one of the methods, tools or procedures for enforcing a security policy designed to prevent and detect any threats that can happen in the system (Security Mechanisms, n.d.). There are various different types of computer security mechanism: physical security, network security, antivirus software, access control, encryption and

Intrusion Detection and Prevention Systems (IDPS). Physical security involves a person who has access to the computer in the location and uses the controls. Passwords, hidden files, and other safeguards cannot be protected with the utmost security. A firewall is a utility or device that will screen data getting into the system or leaving any computer. Access control is access to protected data, and it must be limited to persons who are allowed to access the information Cisco Press (February, 2016). Access control is generally considered in three steps: identification, authentication, and authorization. If the data is encrypted in another form or code, then only those people who have the secret key (decryption key) will be able to access the data to read it. The *ciphertext* is the text that has been encrypted; similarly, unencrypted data can be called *plaintext*. An Intrusion Detection System (IDS) is used to inspect all outbound and inbound network activities and finds suspicious patterns that may point to a network or system attack from an anonymous person attempting to hack into a system. An Intrusion Prevention System (IPS) can dynamically stop traffic by adding regulations to a firewall. An IPS is an IDS that has the potential to detect and prevent attacks (Cisco Press, 2016). Research indicates that DNA can be considered for ultra-scale computation and storage.

The real task comes in securing the data with utmost care to prevent any breach in the system or any network. As there is an increase every year in hacking and other unauthorized activities, it is really challenging for the people who are tasked with trying to find all those illegal intruders in the system. Various attacks could cause harm to the system, yet the mechanism offers few solutions designed to discover any attacks.

The principle of a classification of taxonomy is to *identify* the possibilities of different types of attacks (Hansman and Hunt, 2005). Right now, many of the attacks are generally understood in a different way by each organization, as a result, confusion occurs as to what effect an attack is actually intended to have on DNA computing. For example, an organization "A" may regard an attack as a virus while another organization "B" can classify it as a worm. Here, a precise classification scheme is summarized that can be passed among many organizations. The proposed taxonomy aims to lay out categories in DNA computing so that any attacks can be identified before they can happen as well as to remove those attacks when they enter the DNA system.

A summarized workflow of the work is as follows:

1. Make a study of DNA computing and its associated terminologies
2. Describe the important criteria, like data confidentiality, data integrity, and data availability (CIA), that the security system must offer if the DNA computation and storage are to be trustworthy
3. Classify the list of existing attacks that are causing damage to the user's security, making the user's DNA lose or corrupt his data
4. Offer solutions to prevent as well as detect security attacks on DNA computing by intruders or attackers

In this work, the authors discuss DNA computing and the distinct types of security attacks and how to eliminate these attacks. The rest of the work is organized as follows: The first section contains an introduction. The second section examines the significance of data security. The third section addresses attacks on DNA security, classified as either *identity* or *non-identity*. The last section contains the conclusion and discusses avenues for future research.

6.2 The Significance of Data Security

The term *computer security* refers to the process of safeguarding data from intruders and people who do not have authorized access to the system but who could get into it in order to change or delete data or carry out any modifications that could cause damage to the system. There are three major objectives that need to be achieved in computer security so that it can be ensured that information is not available to unauthorized individuals, changed from the original form, or denied to authenticated users; the three aims are stated in Figure 6.1.

All security measures should satisfy the following three conditions.

1. Retain the confidentiality of data
2. Retain the integrity of data
3. Ensure the availability of data for authorized use

6.2.1 Confidentiality

The main objective of security is *confidentiality*: keeping information away from persons who are not supposed to have it. Confidentiality keeps information from being compromised by an unauthorized user. Achieving this objective ensures that information is secure, and it determines who has the right to use it. Thus, a security system must have a tough mechanism to defend against attackers so that it cannot be breached by them whenever information is exchanged.

In industry, the banking and healthcare sectors, and many other places, information security is critical. Encryption is the standard means of protecting information by transforming it from normal text to unreadable text. Security protocols such as SSL/TLSa are used with some internet protocols to protect data from attack by intruders. Thus, in DNA computing, confidentiality can be achieved by protecting the DNA sequence, which is arranged to facilitate computation or storage that are used to store huge amounts of data, replacing traditional servers and other storage devices. Two different persons interact with an encrypted e-mail by knowing the both encrypted as well as decrypted keys. In a case where an intruder is able to crack the secret key, then the confidentiality of that particular data is lost.

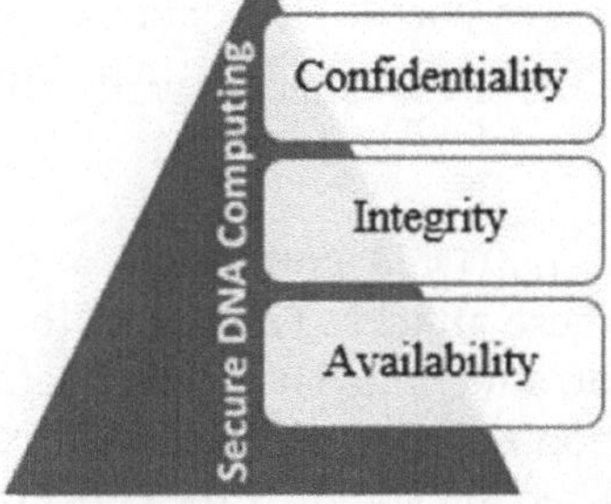

FIGURE 6.1
CIA on DNA computing.

6.2.2 Integrity

The next aspect of security after confidentiality is *integrity*, which means that information should never be modified or deleted from the source. Integrity involves both confidentiality and availability. Data must reach the trusted party and be kept away from intruders. Despite these security solutions, many security failures can still happen the DNA computing as intruders become more skilled at carrying out attacks. Sometimes, the people whom we trust are not trustworthy. Integrity should not be limited to "WH" terms such as *whom* and *what conditions*.

Integrity is maintained in DNA computing when the DNA sequence (Adenine must be combined with Thymine and Cytosine must combine with Guanine) remains intact; this is the sequence that should not be changed by any unauthorized individuals like intruders and attackers. Integrity will be lost if the DNA sequence has been changed in the molecules of the DNA that performs the computation. If someone spends 1,000 of a currency on an online purchase but is shown to have spent 1,000,000, this represents a big failure in integrity.

6.2.3 Availability

The last objective of security is *availability*: ensuring that whoever has the right to access certain data can get them at any time by giving the valid credentials required by the system. Availability can also protect the data against a Denial of Service (DOS) attack, where the attackers are trying to block the data for the intended users.

Distributed Denial of Service (DDOS) attacks block regular users of a site from getting access to data. The resulting downtime is very costly. *Downtime* is the time during which the machine is not able to work. There are other factors that may lead to the unavailability of significant data, such as power outages or natural disasters such as floods, earthquakes, and so on.

DNA is used to store and retrieve the data as well as perform computations for solving a NP problem. Deletion of DNA or any attacks that cause the user to be unable to access the data in DNA will result in the failure of availability. For example, when an intruder has taken control over the server of a bank and destroyed its processing capability, even an authenticated user will not be able to access the e-banking application, leading to resources being unavailable to customers.

6.3 Security Attacks in DNA Computing

In computer networks, an attack is anything that tries to change, stop, destroy, take, or gain unauthorized access to an asset with or without the knowledge of the user of the system. An attack can steal or change the data that are stored in a system, which leads to data loss and many consequences. Intruders carry out such attacks in order to destroy the organization or other entity.

Attacks can be categorized as

1. Identity-based security attacks
2. Non-identity-based security attacks

6.3.1 Identity-Based Security Attacks

In such an attack, the intruder will get access to or is allowed to enter the system by giving some false identity and causing a fault in the system or the network. When DNA computing faces such attacks, like spoofing, key attacks, password attacks, and so on, the attacker or the intruder use the authenticated credentials of the trusted parties, which they will get access to by carrying out those attacks. Each attack has the unique style of compromising the system and attaining access to the main server or the network so that the intruders will be able to complete their intended task; for example, if one intruder gets access to the server of a hospital, then all the details of the patients are in their hands. The intruder can demand anything in return for getting the system back up and running. The DNA computation is performed by the bimolecular reaction, which can run parallel computations simultaneously. Table 6.1 shows comparison between identity-based and non-identity-based security attacks.

It can be compensated by any *security* attacks that are happening in the conventional computing (Figure 6.2).

Identity-based attacks are categorized as

1. Spoofing
   a. IP spoofing
   b. MAC spoofing
   c. Bio spoofing
2. Compromised key attack
3. Active attack
   a. Masquerade
   b. Replay
   c. Modification of messages

TABLE 6.1
Comparison between Identity-Based and Non-Identity-Based Security Attacks

S.No	Category of Attacks	Attack	Explanation
1	Identity-based attacks	IP spoofing	Using a fake IP address
2		MAC spoofing	Using a fake MAC address
3		Bio spoofing	Using DNA genes
4		Compromised key attack	Cracking the secret key
5		Masquerade	Sending false DNA data
6		Replay	Delaying the DNA contents
7		Modification of messages	Changing the DNA data
8		Brute-force attack	Tools generating n number of passwords
9		Dictionary attack	Cracking the password by the familiar words
10		Key-logger attack	Cracking the password by the keystroke
11		Release of message contents	Accessing the DNA data
12	Non-identity-based attacks	Traffic analysis	Observing the pattern of DNA
13		Eavesdropping	Listening between the two systems
14		DDOS	Producing traffic
15		Sniffing	Obtaining credentials
16		Backdoor	Planting a code for intruders
17		Man-in-the-middle attack	Inserting malicious code
18		Application-layer attack	Attacking the DNA computer application

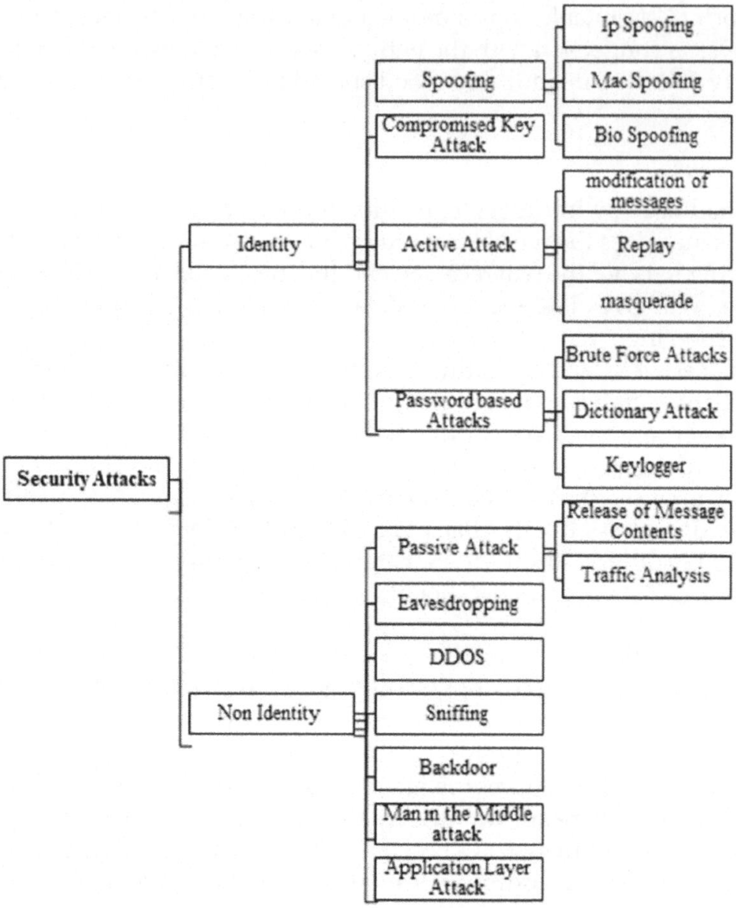

FIGURE 6.2
Classification of security attacks.

4. Password-based attacks
   a. Brute-force attack
   b. Dictionary attack
   c. Key-logger attack

6.3.1.1 Spoofing

This involves giving false information or credentials; the receiver on other end assumes the credentials are genuine, and the intruder gets the information they need.

There are three types of spoofing: Bio spoofing, IP spoofing, and MAC spoofing.

6.3.1.1.1 Bio Spoofing

In a human context, *bio spoofing* means replicating another individual's biological or biometric information. Genetic impersonation is the process of accessing the individual data by using the genetics of the specific individual. Spoofing is the process of collecting DNA in one location and obtaining its biometric traits. With synthetic DNA technologies, it is

possible to "print" acceptable copies of a genome using the previous data, avoiding the need for a physical connection with the entity. This can be transferred into the blood cells of a third party and the contaminated blood spread to the attacker. In this way, people can be framed.

6.3.1.1.2 IP Spoofing

This is a method used by intruders to get unauthorized access to computers, wherein the intruder sends messages to an DNA computer, but the DNA computer is not aware the IP that it is replying to is not the real receiver, and that the intruder is replicating the IP. Then the DNS server will give all the required data, thinking they are authorized, whereas the intruders are spoofing.

Most of our networks and operating systems (OPs) use the IP address of a computer or laptop to identify valid users, but the problem arises when an IP address is faked by an intruder to get into the access of the main network; if this happens, the receiver side will think the information is being sent from the trusted party. There are two different computers that are used to store the data in DNA. Each system, DNA A and DNA B, will have a unique IP address like 192.168.0.25 and 100.0.0.75. The intruder, C, will be faking the receiver B's IP address, 100.0.0.75, so that the intruder has access to the sender A that is being used to store data in DNA, as shown in Figure 6.3.

6.3.1.1.3 MAC Spoofing

This is the process of changing a predefined Media Access Control (MAC) address of a system from an old original one to a new one (even a bank server), so that the information that is sent to the trusted MAC address can be compromised when the spoofing is carrried out. Even though the MAC address is attached to a Network Interface Controller (NIC) that is hard to modify, there are still drivers that permit the alteration of MAC addresses; this masking of MAC addresses is known as *MAC spoofing*. This type of attack is similar to IP spoofing where IP addresses are faked by the intruder, but in MAC spoofing, when the DNA computation is executed in a computer A, which has the MAC address of 00:00:00:a1:2b:cc, and a situation arises in which an intruder is able to get the MAC address (00:00:00:a1:2b:cc) of User A, then the computations for solving a problem that are performed by the DNA can be changed by the intruder, as shown in Figure 6.4.

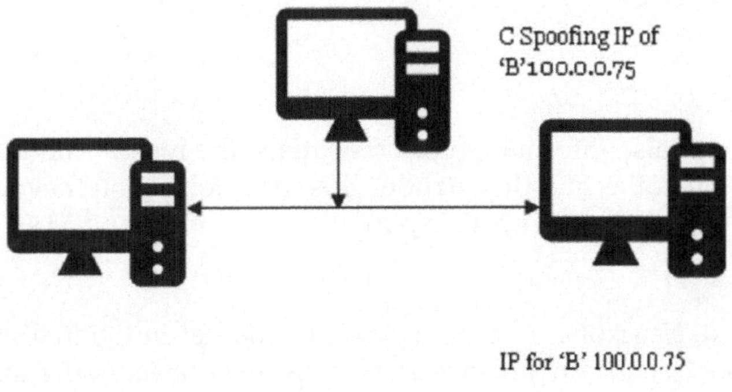

FIGURE 6.3
IP spoofing in DNA computing.

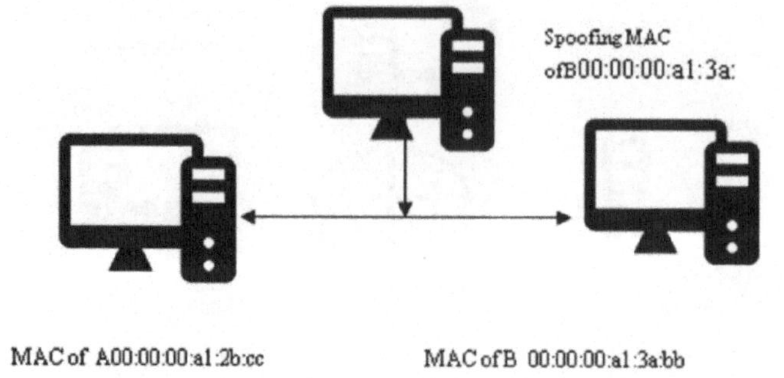

FIGURE 6.4
MAC spoofing.

6.3.1.2 Compromised Key Attack

An intruder can cause this attack when they get the secret key that is used in the process of encrypting as well as decrypting the information. Even a malware can be encoded in DNA and used to attack vulnerabilities in the computer. In simple terms, these are the most common attacks that can be done by intruders to get the privileges of the server or any network system that can store the most sensitive information pertaining to whole industries, institutes, banks, and so on. Even though finding the secret key is not easy, it is possible to crack such a key and this is known as *compromised key* (Common Types of Network Attacks. [n.d.]. Retrieved 2018).

For example, when sender A sends some files that are stored in the form of DNA and they are transferred to system 'B' by using the encryption method as a secret key, an intruder can break the encryption by using these attacks and can steal the information, resulting in the loss of the confidentiality of the DNA, as shown in Figure 6.5.

6.3.1.3 Active Attack

In *active attacks,* the intruder listens to the DNA system by hacking the network and changes the data of the DNA that are stored in it by introducing inaccurate and faulty DNA into the system.

It is classified into different types such as masquerade, replay, modification of messages, and denial of service.

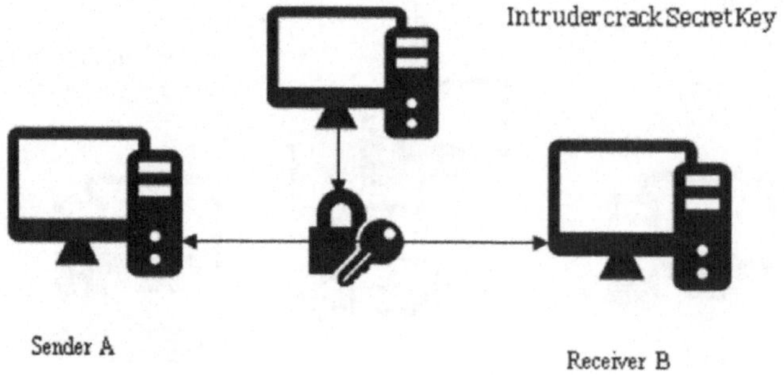

FIGURE 6.6
Compromised key attack.

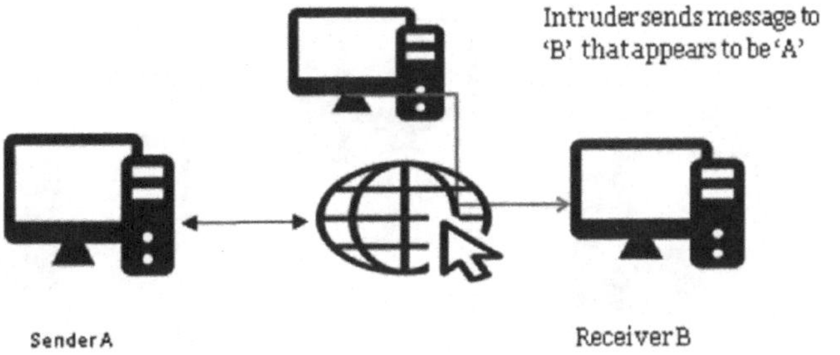

FIGURE 6.6
Masquerade.

6.3.1.3.1 Masquerade

In a *masquerade* attack, hackers impersonate the trusted individuals who are authorized to access and use the system (Copyright and Stallings, 2013).

For example, if there are three different DNA systems, A, B, and C, B (intruder) will send a message through DNA to C (receiver), but C will think that the DNA message was from A (trusted sender), whereas A was not involved in the sending of any of DNA-based messages, as shown in in Figure 6.6.

6.3.1.3.2 Replay

This refers to a network attack where the intruder does not alter the messages but the messages can be delayed or repeated; thus, the intruder may get access to the message and then it will reach its destination, where the receiver remains unaware that any attack has happened (Copyright and Stallings, 2013).

Take, for example, a case where there are three different systems performing DNA computation, sharing, and storing of messages or files: A, B, and C. When DNA system A sends a DNA message to DNA system C, B (intruder) will compromise the information from A and will then pass this particular DNA message on to C, as shown in Figure 6.7.

6.3.1.3.3 Modification of Messages

This involves the text of the message being altered or messages being suspended or not in sequence, an attack in the system that is called *modification of messages* (Copyright and Stallings, 2013).

For example, DNA system A sends a message to DNA system B meaning "Allow Rahul to read confidential file accounts," and this is altered by the intruder to mean "Allow Ravi to read confidential file accounts," as shown in Figure 6.8.

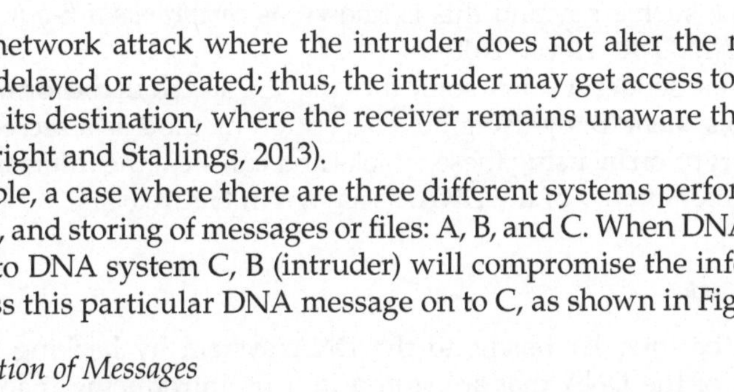

FIGURE 6.7
Replay.

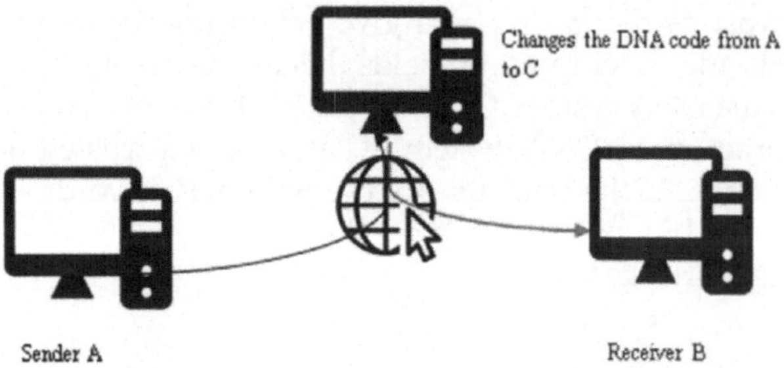

FIGURE 6.8
Modification of messages.

6.3.1.4 Password-Based Attacks

This is an attack in which *n* number of attempts will be made by the attacker to find valid credentials for the DNA system. *Authentication* is the process of confirming the identity of a particular user of the DNA system by giving valid credentials such as passwords, biometrics, and smart cards, but the most familiar authentication mechanism is password-based authentication, so attackers or intruders will use these mechanisms to get into the DNA system (3 Types of Password Security Attacks Retrieved, 2018). Even computers that are used to perform DNA computation for solving problems use passwords as a way to authenticate the user, as explained in Figure 6.9.

There are few steps that can be taken to overcome password-based attacks:

1. The password should not be revealed to anyone.
2. Individuals should avoid using the vendor default password (like User12345).
3. Passwords should be used that are impossible to crack.
4. Passwords should not be stored in online locations like e-mail.
5. Passwords can be combined with multifactor authentication for more security.

The following are familiar types of password-based attacks:

1. Brute-force attack
2. Dictionary attack
3. Key-logger attack

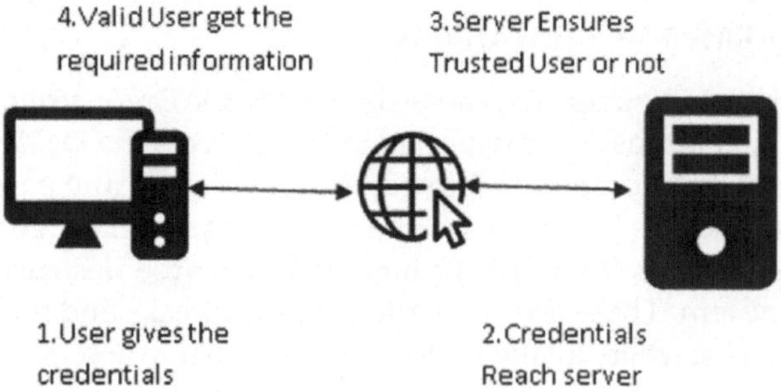

FIGURE 6.9
Password-based attack.

The above flow explains that a user will give valid information to the server, such as username, passwords and other required fields that are necessary to complete the authentication to enter into the DNA system. Once the details have been confirmed by the respective server or any other main DNA system, whether it is a trusted one or not, then the requested data can be sent to the DNA user of the system. Passwords are the most familiar way to log into a system, but intruders can get the upper hand by making brute-force, dictionary, and key-logger attacks.

6.3.1.4.1 Brute-Force Attack

This is the process of making attempt to predict a password till a successful login happens (Ahmed and El-Henawy, 2017). It is a kind of trial-and-error method, where applications are automated to find the most probable passwords that the user might use to authenticate the DNA system. These types of attacks can be prevented by making passwords much tough to predict; credentials must be greater than six characters and must have a combination of alphanumeric characters and special symbols. Avoid using dates of birth, names, mobile numbers, and so on.

6.3.1.4.2 Dictionary Attack

This uses the most familiar and frequent words to access the credentials of the user. It works on many computers within organizations, as those DNA computers will use the most familiar words as passwords. These attacks are self-automated and there are many tools the internet for performing this type of attack on the DNA system. It can be prevented by locking the account after three unsuccessful attacks on the DNA computers. These attacks are most successful when the DNA user has mostly familiar words as passwords, but the success rate is less when the user credentials are a combination of uppercase and lowercase letters, numbers, and special symbols as those are not in the application.

6.3.1.4.3 Key-logger Attack

This type of attack involves storing a record of every keystroke that the user has typed on DNA computer keyboard. Key logger is a most powerful threat, compromising any person's login access, such as username and password, and attacking the DNA system by using the credentials attained in the key-logger attack.

6.3.2 Non-Identity-Based Security Attacks

In this type of attack, the intruder will cause damage to the DNA system or the network by attacking the DNA without using any false identity to access the DNA system.

In these attacks on DNA computers, intruders will not need any password, credentials, or any authentication details belonging to the trusted party and they will be still able to compromise the network or the DNA system and cause the destruction of data on the server or the DNA system. These are also called *indirect* attacks and represent the opposite of identity-based attacks, where a fake identity will be used to access or change the data in a DNA system. DNA users are generally aware of the identity-based attacks that have happened, whereas users are not aware of those attacks on the DNA system in non-identity-based attacks.

These include

1. Passive attacks
   a. Release of message contents
   b. Traffic analysis
2. Eavesdropping
3. DDOS
4. Sniffing
5. Backdoor
6. Man-in-the-middle attack
7. Application-layer attack

6.3.2.1 Passive Attacks

In this type of attack, the DNA data are monitored, the ultimate motive of the attacker being to determine the DNA data that are sent to the receiver end without changing the contents of the DNA data.

There are two different types of passive attack:

1. The release of message contents
2. Traffic analysis

6.3.2.1.1 Release of Message Contents

This is the process of listening to the interaction between the DNA data sender and DNA data receiver without the knowledge of either sender or receiver (Copyright and Stallings, 2013). Real-time scenarios like telephone conversations, mail communication, and chatting on social media can include confidential information. The attacker must be prevented from knowing the contents of the DNA message by implementing strong access-control mechanisms in DNA computers, as explained in Figure 6.10.

6.3.2.1.2 Traffic Analysis

Traffic analysis is similar to release of message contents, but whereas in the latter attack, the contents of the DNA computer A can be read by the intruder, in traffic analysis, it is not possible to read the content of the DNA computer A, but the intruder can access the content pattern that is sent from system A to B (Copyright and Stallings, 2013). Encryption is

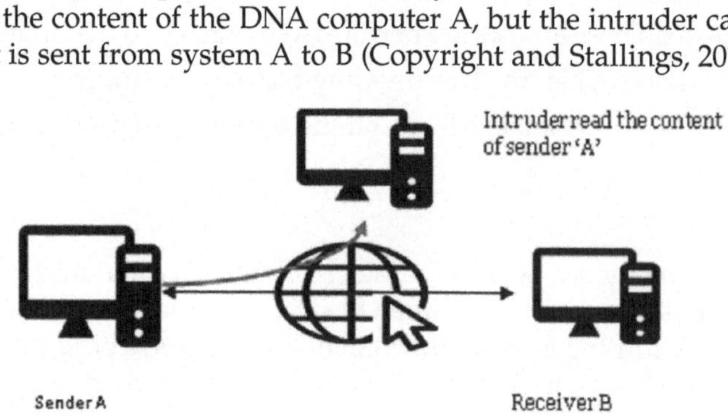

FIGURE 6.10
Release of message contents.

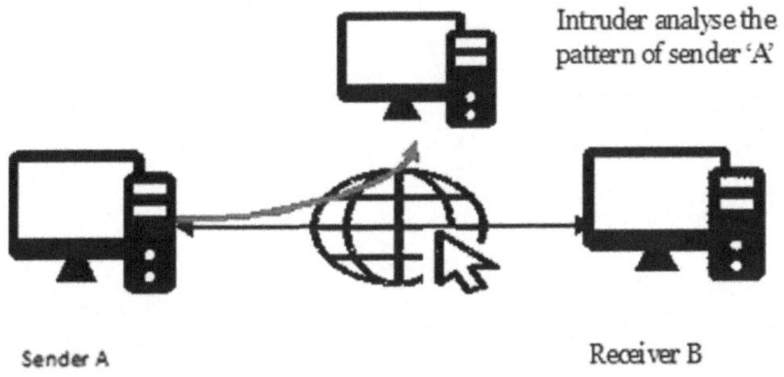

FIGURE 6.11
Traffic analysis.

the standard way to protect the information; here the intruder can find the pattern that the sender DNA system, A, uses to communicate with the receiver DNA system C, as shown in Figure 6.11. This information can help the attacker determine what type of data was sent and received by the sender and the receiver.

6.3.2.2 Eavesdropping

An eavesdropping attack can take advantage of network communications that are not secured (Jung, 2010), thereby accessing the DNA data being sent and received. Eavesdropping attacks are hard to detect since they do not cause any decline in DNA computation performance or changes in the DNA data from the original form. This is also known as a *sniffing* or *snooping* attack.

6.3.2.3 DDOS

This type of attack involves sending millions of units of traffic to one DNA computer to cause the DNA machine to slow down, while disabling a certain security feature; meanwhile, the attackers do their DNA data stealing from the server (Lesk, 2007). A *bot* is the computing device that is in the hands of the intruder; commanding bots are called *botnets*, which are used to perform DDoS attacks, as explained in Figure 6.12.

There are four different attacks that can be performed in this category.

1. *TCP connection attacks*: These attacks are used to occupy connections.
2. *Volumetric attacks*: These attacks are performed by using up maximum bandwidth.
3. *Fragmentation attacks*: This involves dividing packets into pieces.
4. *Application attacks*: These attacks focus on the computer's applications.

6.3.2.4 Sniffing

In parallel with the emergence of the ethernet, sniffers also came into being, which have the potential to see data transmitted between the DNA data sender and DNA data receiver. The objective of DNA sniffing is to obtain credentials like passwords e-mail text, and files in transfer (Types of Attacks. [n.d.] 2018). Thus, when the attacker uses a sniffer, any one of the following can happen (Common Types of Network Attacks. [n.d.)]. Retrieved 2018) (Figure 6.13). The sniffer can

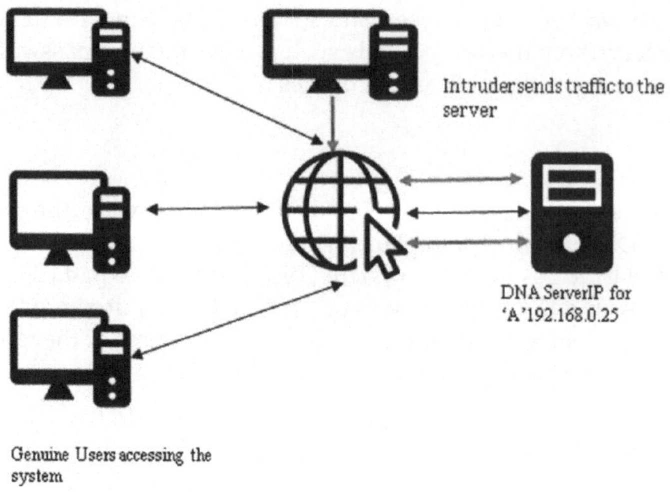

FIGURE 6.12
DDOS.

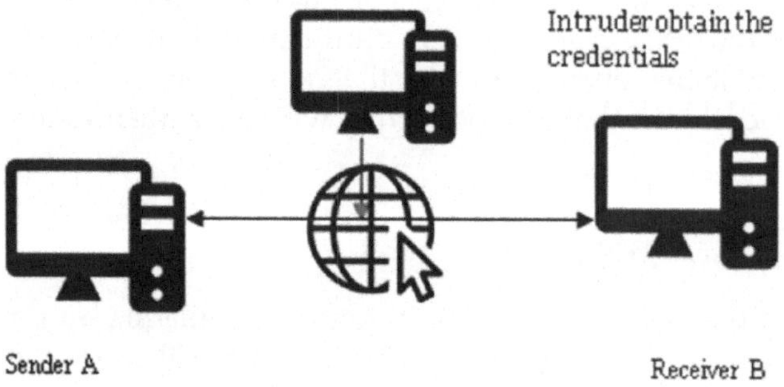

FIGURE 6.13
Sniffing.

- Inspect the DNA network and acquire the DNA data to change DNA network as per the intruder's wish
- Monitor and see communication between the DNA data sender and the DNA data receiver.

Types of Sniffer Attacks

1. *LAN sniffing*: The whole DNA network can be monitored by the sniffer once the software has been installed on the internal local area network (LAN).
2. *Protocol sniffing*: Separate and specific sniffers are used to perform attacks on different protocols.
3. *ARP sniffing*: Through these attacks, the intruder will be able to access data about the IP and MAC addresses of the DNA computers that perform computations.
4. *TCP session sniffing*: Here, the intruder gets access to the source and the destination IP address where DNA message transmission occurs between the two DNA systems.

5. *Web password sniffing*: The hacker attacks the web using HTTP sessions that won't use encryption; then credentials such as user IDs and passwords are easily obtained for performing malicious activities.

6.3.2.5 Backdoor

A method that allows the attacker to get around all the regular ways, the backdoor involves getting access to a DNA network and planting a program that creates a space for an unauthorized individual to access the DNA system (Types of Attacks. [n.d.] 2018). Backdoor lets the DNA user access the DNA system without giving the required credentials, such as a password. Back orifice netdevil, subs even, and netbus are some of the tools used to carry out backdoor attacks on DNA computers.

6.3.2.6 Man-in-the-Middle Attack

Whenever DNA user 1 and DNA user 2 communicate with each other and this has been actively monitored, captured, and controlled by the intruder, then it is called a *man-in-the-middle attack* (Man-in-the-Middle *(MITM)* Attack, 2017).

The attack involves inserting a malicious code between the DNA server and the authenticated user. Either the DNA user or server administrator may be aware of the attack, but the software conveys to the server or the user that no suspicious activities were carried out. thus, the DNA server thinks that it is interacting with the authenticated user, as shown in Figure 6.14.

6.3.2.7 Application-Layer Attack

This attack targets the DNA server's application by simulating an error in the server's operating system, as shown in Figure 6.15. The intruder will attack the DNA application used to make computations, and once the intruder has succeeded in making the attack, then the user is no longer able to access the server to perform the computation (Common Types of Network Attacks.[n.d.]. Retrieved 2018).

The operations that can be done by attacker are as follows:

- Reading, adding, deleting, or modifying user's DNA data or operating system
- Sending a virus to the DNA network that will replicate *n* number of viruses into the DNA network

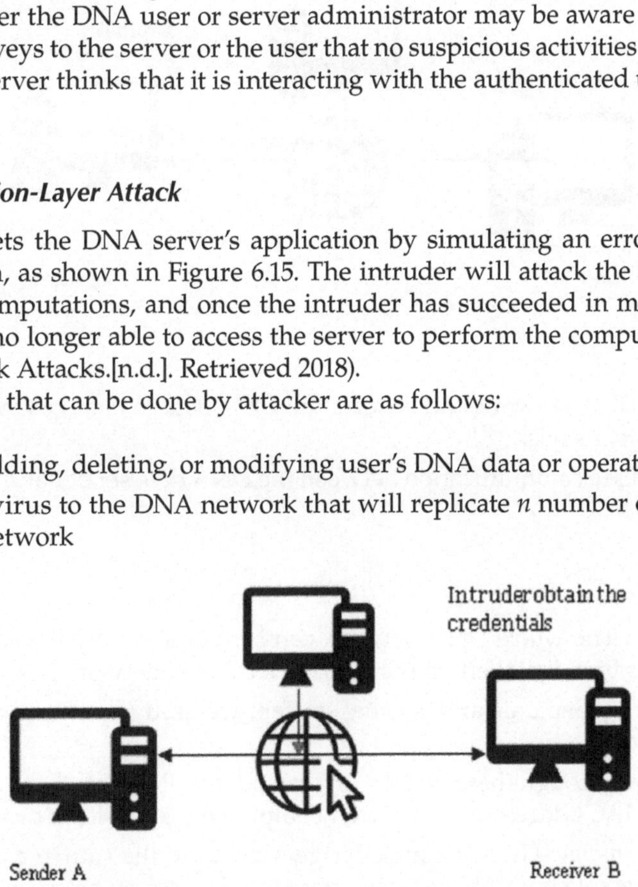

FIGURE 6.14
Man-in-the-middle attack.

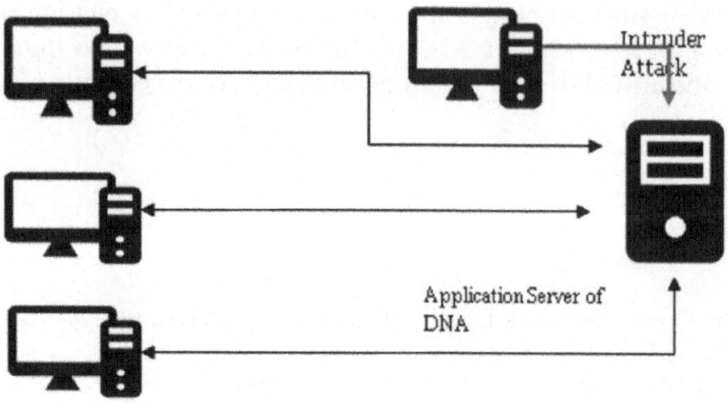

FIGURE 6.15
Application-layer attack.

- Suddenly terminating user's data applications or operating systems
- Making other security measures inactive in order to enable future security attacks

For example, three different users are trying to access the DNA server application, but the intruder has attacked the server and the others will not be able to access the application in order to perform any complex computations.

6.4 Conclusion

DNA computing is an emerging technology for the storage of data and computation of complex problems that are not easily solvable by conventional computing, and it can sometimes even act as a *security* mechanism along with cryptography and so on. But the disadvantage is that there are *n* number of attacks that can be performed by an intruder on the DNA computer, which results in the loss of DNA confidentiality, DNA integrity, and DNA availability. This chapter has summarized the most common types of attacks that can happen on a DNA network or a DNA system and how to defend against these security attacks. The classification of identity and non-identity-based attacks helps to understand clearly the causes and the effects of those attacks by specifying the reasons why attacks can happen and the intruders who make the attacks. The taxonomy of security attacks on DNA computing can be handy when DNA computers face similar attacks, so that storage and computation can be done on DNA computers without suffering any attacks.

Key Terms and Definitions

DNA computing: This involves the execution of computations by biological molecules instead of silicon chips. This emerging computing technology converts the data in a particular problem to DNA oligonucleotides. Biomolecule tools are used for the decryption of DNA that is encrypted to obtain computation results.

Security attacks: With such an attack, the intruder can steal or change the data that are stored in a system, which leads to data loss and many consequences. It is carried out with the aim of destroying an organization or other entity.

References

Adleman, L. (1995). Computing with DNA. *Journal of Computational Biology*, 2(1), 1–7. https://doi.org/10.1089/cmb.1995.2.1

Ahmed, K. and El-Henawy, I. (2017). Increasing robustness of data encryption standard by integrating DNA cryptography. *International Journal of Computers and Applications*, 39(2), 91–105. https://doi.org/10.1080/1206212X.2017.1289690

Brent Cornell. Retrieved February 24, 2018, from http://ib.bioninja.com.au/standard-level/topic-2-molecular-biology/26-structure-of-dna-and-rna/dna-structure.html

Cisco Press (2016, February 9). Retrieved February 24, 2018, from http://www.ciscopress.com/articles/article.asp?p=1626588&seqNum=2

Common Types of Network Attacks (n.d.). Retrieved February 24, 2018, from https://technet.microsoft.com/en-us/library/cc959354.aspx

Cui, G., Qin, L., Wang, Y., and Zhang, X. (2007). Information security technology based on DNA computing. *2007 IEEE International Workshop on Anti-Counterfeiting, Security, Identification*, Tianjin, China, 288–291. https://doi.org/10.1109/IWASID.2007.373746

Cui, G., Li, C., Li, H., and Li, X. (2009). DNA computing and its application to information security field. *2009 Fifth International Conference on Natural Computation*, 6, 148–152. https://doi.org/10.1109/ICNC.2009.27

Gehani, A., LaBean, T., and Reif, J. (2003). DNA-based cryptography. *Aspects of Molecular Computing*, 2950, 167–188. https://doi.org/10.1007/b94864

Hansman, S. and Hunt, R. (2005). A taxonomy of network and computer attacks. *Computers and Security*, 24(1), 31–43.

Jung, E. J. (2010). Privacy and security definition of security/privacy attacks, services and mechanisms passive attack (1) – eavesdrop passive attack (2) – analysis.

Lesk, M. (2007). The new front line: Estonia under cyberassault. *IEEE Security and Privacy Magazine*, 5(4), 76–79. https://doi.org/10.1109/MSP.2007.98

Man in the Middle (MITM) Attack. (2017, August 15). Retrieved February 24, 2018, from https://www.veracode.com/security/man-middle-attack

Ning, K. (2009). A pseudo-DNA cryptography method. *09032693*. https://doi.org/10.1016/j.compeleceng.2012.02.007

Saghir, H. and Megherbi, D. B. (2015). Big data biology-based predictive models via DNA-metagenomics binning for WMD events applications. *2015 IEEE International Symposium on Technologies for Homeland Security, HST 2015*, IEEE, Waltham, MA, 2–7. https://doi.org/10.1109/THS.2015.7225313

Security Mechanisms (n.d.). Retrieved February 24, 2018, from https://www.rbc.com/privacysecurity/ca/security-mechanisms.html

Stallings, W. (2013). *Cryptography and Network Security*.

Types of Attacks. (n.d.). Retrieved February 24, 2018, from https://www.go4expert.com/articles/types-of-attacks-t7685/

Types of Password Security Attacks and How to Avoid Them. (n.d.). Retrieved February 24, 2018, from https://authanvil.com/blog/3-types-of-password-security-attacks-and-how-to-avoid-them

UbaidurRahman, N. H., Balamurugan, C., and Mariappan, R. (2015). A novel DNA computing based encryption and decryption algorithm. *Procedia Computer Science*, 46(Icict 2014), 463–475. https://doi.org/10.1016/j.procs.2015.02.045

Wang, X. and Zhang, Q. (2009). DNA computing-based cryptography. *BIC-TA 2009 – Proceedings, 2009 4th International Conference on Bio-Inspired Computing: Theories and Applications*, 67–69. https://doi.org/10.1109/BICTA.2009.5338153

7

Security, Privacy, Trust, and Anonymity

Suyel Namasudra, Debashree Devi, Sandeep Choudhary, Rizwan Patan, and Suresh Kallam

CONTENTS

7.1 Introduction

Information or data are the core of any organization or system. For effective functioning of the organization or system, the information must be secured against any kind of attack or breach. Nowadays, network-based services have taken over the internet, generating huge amounts of information at any one time. The processing of such huge amounts of data requires the employment of intelligent tactics, which can provide rapid accessing in a secure way. Defining an efficient security model involves various objectives that need to be satisfied, namely data confidentiality, data integrity and data availability, known as the *CIA triad* for short. The CIA triad helps users to develop security measures or policies that can offer strong protection for sensitive data without hindering organizational activities. Today, a lot of money is spent on maintaining information databases as they contain a lot of sensitive information relating to such things as personal data, legal documents, banking details and national defense programs. A breach in such an information system can allow hackers or attackers to gain access to all those data in a single blow. Hence, it is essential for any organization to implement strong security policies or security measures to protect its data from any kind of attack or security breach.

Data security brings the concepts of privacy and trust to any security model. The aspects of privacy and trust arise due to increasing doubt about protection of users' sensitive data that are shared over the network. To be precise, data protection identifies what information should be shared without releasing sensitive personal data. The accessibility of huge quantities of databases storing an extensive assortment of data about people makes it possible to find data about particular people by basically associating all the accessible databases.

Protecting privacy of data on the internet involves protecting anonymity. The term *anonymity* means "namelessness." Anonymity defines a situation, whereby the real identity of data or of an individual is hidden in order to protect it from unauthorized access. Anonymity allows the realization of other factors, such as privacy or authorization. Anonymity furnishes data security as the owner of a data remains unidentifiable throughout a transmission or transaction. Anonymity is also seen in real-life situations, such as crime investigation, charity donations and defense programs. However, it is sometimes illegal to be "anonymous," when it comes to such serious situations as terrorist activities, etc.

Anonymity is implemented in information security to ensure privacy and the protection of an individual's real identity. In an information system, a lot of information is stored and shared across a number of databases. Since the database contains sensitive information, the fundamental concern is to ensure the protection of users' sensitive data. Anonymity contribute to this goal by making the database nameless. Naturally, if the database is unknown, it is not conceivable that the users' identities could be deduced from the data contained in the database.

DNA computing is one of the trending fields in biology and computer science. DNA computing provides a tool for parallel computing, and it takes advantage of the numerous different molecules of DNA (Lewin, 2002). The complexity and organization of all human beings is mainly based on the coding system of the four components of the DNA molecule. One strand of DNA consists of four bases (A, T, C and G). Two strands of a DNA molecule are linked in a double-helix structure and are anti-parallel, i.e., the flow of information in the strands is opposite to each other. The foundation of this double-helix structure is the bonding between the DNA bases, i.e., A with T and C with G. Figure 7.1 depicts the double-helix structure of a DNA molecule.

After attaching themselves to the deoxyribose, those nucleotides can together form a string for generating long sequences (Boneh and Lipton, 1996).

Different DNA structures are used by researchers for solving different problems (Lipton, 1995). DNA was first proposed in the field of computation by Leonard Adleman in 1994 at the University of Southern California. Adleman demonstrated a proof-of-concept by using DNA in the form of a computation that solved the seven-point "Hamiltonian Path Problem" (Adleman, 1994). Initially, this novel approach was used to solve NP-hard problems. However, very soon it was realized that DNA computing may not be the best solution for this type of problem. Computer scientist Mitsunori Ogihara and biologist Animesh Ray described an implementation of Boolean circuits by using DNA computing in 1997 (Ogihara and Ray, 1999). Since then, many techniques have been developed by using DNA to solve many problems, such as the SAT problem, the 0-1 planning problem, the integer planning problem, graph theory, optimal problem, database, cryptography, etc., and many Turing machines have been proved. The benefits of DNA computing are:

1. Speed
2. Minimal storage requirements
3. Minimal power requirements

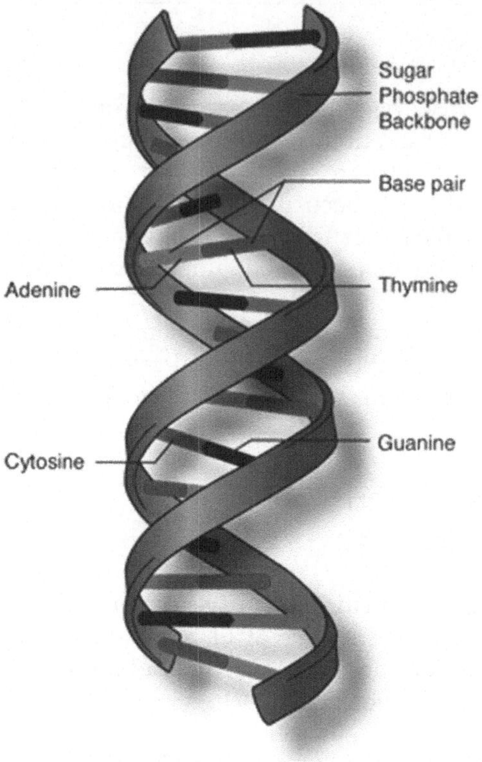

FIGURE 7.1
Double-helix structure of DNA molecule.

Moving on to security concerns, *DNA cryptography* is a trend that has recently emerged, which has been developed based on DNA computing (Tornea and Borda, 2009). Here, DNA computing is used for data encryption and data storage to create a strong security mechanism, so that unauthorized malicious users and attackers are unable to read the data content. DNA computing is implemented to challenge the conventional cryptosystem by facilitating parallel computing to solve different parts of a computing problem, a method that is more secure than conventional public-key cryptography. Many cryptographic algorithms have been proposed based on DNA, namely symmetric and asymmetric key cryptosystems, triple-stage DNA cryptography, DNA steganography systems, DNA-based chaotic computing and DNA-based encryption algorithms (Namasudra, 2017, 2018; Namasudra and Roy, 2015, 2016, 2017a,b,c, 2018; Namasudra et al., 2014, 2017a–c; Sarkar et al., 2015). As for the pros of DNA cryptography, it is much more secure than conventional cryptography methods as it is quite difficult to pre-guess the exact ordering of DNA bases. Additionally, a large volume of data can be stored in DNA bases, which reduces the cost of large storage space.

The main contribution of this chapter is to explore the aspects of security, privacy, trust and anonymity in DNA computing. All these aspects have interrelated concepts. However, each one has a particular role in the protection of user data. The implementation of DNA computing in treatment of these aspects is studied thoroughly in this chapter.

7.2 Security

The concept of data security arises as there are numerous hackers or intruders over the internet, who attempt to get unauthorized access to users' sensitive data; this involves modification, distortion, disclosure or disruption of information (Introduction to Information Security, 2018) (Data Security, 2018). Data or information security basically evolved to define privacy measures for preventing unauthorized access through implementing efficient security policies without impeding organizational activities (Data Security, 2018).

In order to define a security model for any system or organization, three main goals need to be achieved: confidentiality, integrity and availability, or the *CIA triad* for short (Figure 7.2). Any security measure should satisfy the following three conditions:

1. Saving the confidentiality of data
2. Retaining the integrity of data
3. Ensuring the availability of data for authorized users.

The basic motivation of information security is to provide a balance between protection and the confidentiality, integrity and availability of data by maintaining an efficient security policy without hindering the productivity of an organization/system. The CIA triad illustrates the basic goals required for achieving information security. The CIA triad furnishes a guide to designing an efficient security policy or model, one that can provide information protection, while satisfying all three objectives of security measures. A brief description of each objective and their role in security measures is presented below.

7.2.1 Confidentiality

Confidentiality forms the basis of any information security model, i.e., the protection of information from unauthorized access. Confidentiality prevents information being compromised by unauthorized persons or sources. Confidentiality ensures that information is secure, and that the lawful owners have the right to use it. Information security models should offer techniques with strong mechanisms to enforce confidentiality for secure data

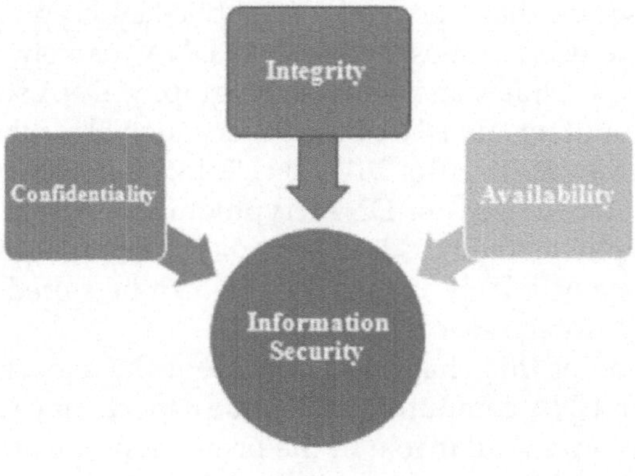

FIGURE 7.2
CIA triad.

transmission. Encryption is a standard means of providing confidentiality of information by means of data encoding, which converts a normal text to a ciphertext. Security protocols, such as SSL/TLS are used with some internet protocols to protect data from hackers or intruders. Implementing file permissions and access control are some alternative approcahes for ensuring inforamtion confidentialty in a security model.

In DNA cryptography, confidentiality can be achieved by protecting the DNA sequence, which is arranged in order to attain the computation or storage from the intruders that are used to store the huge amount of data replacing the traditional servers and another storage device.

7.2.2 Integrity

Integrity refers to the protection of information from being altered by unauthorized users. Integrity of data ensures the transmission of data in its correct form to the rightful person. Integrity is a very important objective for data security as any information is only valuable inasmuch as it is valid. Breaching attacks that target data integrity may cost an individual or organization a lot of money. For example, let us say that a person has carried out a transaction of Rs. 1000/-. But, due to security breaches, the transaction has debited Rs. 1,00,000/- from the account. In the CIA triad, data integrity is maintained, when the information remains unaltered during any transmission, storage or usage. Cryptography provides data integrity along with data confidentiality. A common tactic for maintaining data integrity is *hashing*, where the original values of data are converted to hash (complex) values by using hash functions, which are then used for secure transmission. Access control is another tactic for maintaining data integrity.

The integrity is maintained in DNA computing, when the DNA sequence, in which Adenine is combined with Thymine and Cytosine must combine with Guanine, is maintained; this sequence must not be modified by any unauthorized individuals like intruders and attackers. Integrity will be lost, if the DNA sequence has been changed in the DNA molecules that perform the computation.

7.2.3 Availability

Availability of data allows authorized users to access data whenever they need it, without any hindrance. Any information is valuable, if it is accessible to the right people at right time. Data availability can be maintained, when the system/organization is functioning properly. A common attack targeting data availability is the Denial-of-Service (DoS) attack, where authorized users are prevented from accessing data from a website. Availability of data can be ensured by proper maintenance of hardware, along with implementation of data-specific tools. Some potential solutions for maintaining data availability are given below:

1. Transmitting data through suitable bandwidths
2. Preventing bottleneck situations
3. Implementing Redundancy Array of Independent Disks (*RAID*) to store data in multiple sources
4. Implementing failover, which works as backup operational mode and can assume the functions of a system component by means of secondary component in case of primary component failure

5. Applying redundancy to file/data storage
6. Implementing a fast and adaptive recovery mode in case of entire system failure or natural disaster, such as a flood or earthquake.

Data or information security in cyberspace is very important, as it is essential to protect the information of organizations that have invested a lot of money in Information Technology (IT) infrastructure. Recently, improved technologies open up ways for hackers or intruders to exploit loopholes in data security systems. The threats to data security can take the form of software attacks, identity theft, theft of information, sabotage or extortion of information. *Software attacks* are experienced due to viruses, phishing attacks, etc. *Identity theft* refers to a situation, wherein someone's personal information is stolen in order to access his/her vital information. *Sabotage* defines a scenario, when an organization's website is destroyed with the aim of decreasing user confidence. *Information extortion* refers to a scenario, where information is stolen in order to extort money as a ransom from an organization/system. A pictorial presentation of the various threats is presented in Figure 7.3. The most recent and largest cyberattack was WannaCry, a ransomware cryptoworm that targeted the systems running the Microsoft Windows operating system by demanding ransom money in the form of Bitcoin Cryptocurrency (WannaCryRansomware Attack, 2017).

Irrespective of the motivation behind such attacks, the strategy of data security has to deal with the common elements of people, process and technology (Data Security, 2018). The objectives of data security can differ; they could include protection of a brand, intellectual capital or customer information or perhaps the control of critical infrastructure. Moreover, data security forms a large section of the General Data Protection Regulations (GDPR). Any kind of breach or leak of confidential information may incur a huge fine for the organization; hence, it is necessary to protect against any kind of attack or threat (Malik and Patel, 2016).

A number of techniques are defined as security solutions for information/data security schemes, which are as follows.

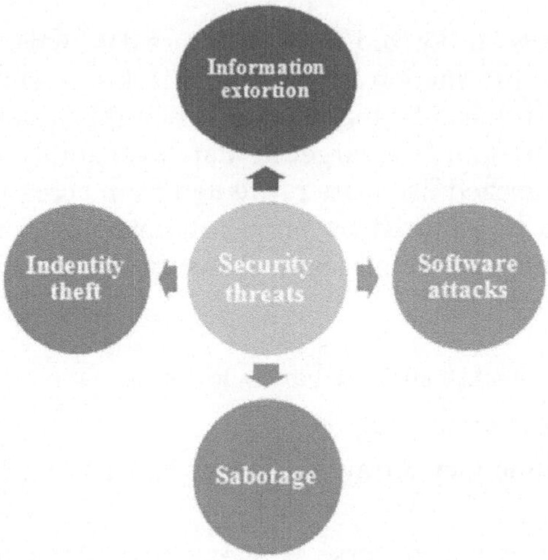

FIGURE 7.3
Various security threats.

7.2.3.1 Data Encryption

Data encryption is a security method that provides encoding of information, making it accessible only to authorized users. Encryption algorithms transform a message into a ciphertext, which can be read by an authorized user after decryption by using a key. Based on the type of key used, data encryption can be of two types:

1. *Symmetric key encryption*: The same cryptographic key is used for data encryption and data decryption.
2. *Public key encryption*: A pair comprising a public and a secret key is used for encryption and decryption. The keys are linked to each other through some mathematical function. The Rivest-Sharmir-Adleman (RSA) algorithm is a popular cryptosystem with public-key cryptography, widely used for secure transmission of sensitive data over the internet.

Hashing is another technique of data encryption, whereby data are encrypted by using a hash function to generate a hash code. The hash code is the encrypted form of a input, which is then used for secure transmission.

In case of data encryption, DNA cryptography uses the concept of cryptography in one-time pad with high computational capability, and incredible compact information storage structure. Introducing DNA cryptography in to common Public Key Infrastructure (PKI), researchers are able to perform data encryption by using DNA-based long public and private keys, which are hard enough to be predictable. One-Time Pad (OTP) encryption uses a codebook of random data to convert plaintext to ciphertext (Nixon, 2002), where the codebook is used as the key. The codebook must be unpredictable with unique data. These two principles, true randomness and single use of pads, dictate certain features of the DNA sequences and sequence libraries, in implementing DNA computing to OTP cryptosystem, at first, a long one-time pad is assembled in the form of a DNA strand, by randomly combining DNA bases, which is then isolated and cloned. These one-time pads are very confidential and are shared only with authentic senders and receiver.

Another DNA cryptography technique is the substitution-based OTP cryptosystem, which converts a plaintext binary message to ciphertext by defining a table with random mapping to ciphertext (Nixon, 2002). The input strand is of length n and is partitioned into plaintext words of fixed length. The table maps all possible plaintext strings of a fixed length to corresponding ciphertext strings, such that there is a unique reverse mapping. Encryption is performed by substituting each plaintext DNA word with a corresponding DNA cipher word. The mapping is implemented using a long DNA pad consisting of many segments, each of which specifies a single plaintext–cipher word mapping. The plaintext word acts as a hybridization site for the binding of a primer, which is then elongated. This results in the formation of a plaintext–ciphertext word pair. Further, cleavage of the word pairs and removal of the plaintext portion must be performed.

7.2.3.2 Data Masking

Data masking allows specific portions of data to be covered, making it unpredictable for hackers or intruders, and thus, protecting the sensitive information. A number of data masking techniques are available, such as substitution, shuffling, number and date variance, nulling out or deletion and masking out. In the case of the substitution technique, the real values of the data are replaced by authentic looking value. In shuffling, the real values

are randomly shuffled within the values of its own attributes. A variance within a range is applied to morph the original data values of number and date. In case of nulling out or deletion, a null value is simply applied to mask the real data values. Masking out involves hiding a certain section of the data field in order to protect the original data. Data masking can be either static or on-the-fly, based on when data masking is applied to the data.

Data masking can be achieved by implementing DNA steganography. DNA steganography was introduced by Carter Bancroft to encrypt hidden messages within microdots during World War II (Nixon, 2002). DNA steganography follows a simple code to convert the letters of the alphabet into combinations of the four DNA bases, and it creates a strand of DNA based on that code. A strand of DNA depicts the message to be masked by encrypting it in the middle of the stand with short marker sequences at both ends of the message. The encoded strand of DNA is then placed into a normal strand of DNA, which is then mixed with DNA strands of similar length. The mixture is then dried onto paper that can be cut up into microdots. Every microdot consists of billions of DNA stands, making it untraceable, and thus, impossible for hackers to determine the original text message.

7.2.3.3 Data Erasure

Data erasure is a software-based method that allows all the data in a device to be wiped, the aim being to destroy all the electronic data using zeros and ones, so that the data cannot be reused in the future.

7.2.3.4 Data Backup

Data backup is the process of archiving data into a source other than the parent source for facilitating restoration in cases of data loss. Data repository models form the basis of any backup system, which can be of various types, such as incremental, differential, unstructured and reverse delta. The backup data can be stored in a storage medium from which it can be accessible in future. Magnetic tapes, hard disks, optical storage, remote backup services, etc. are generally used as the storage medium.

7.2.3.5 Software/Hardware-Based Techniques

Software-based security solutions involve data encryption, whereas hardware-based security can be achieved through security tokens. PKSC#11 is a popular cryptographic standard, which provides two-factor authentication by using tokens as well as PIN authentication. Figure 7.4 presents a pictorial overview of security solutions.

7.3 Privacy and Trust

Trust is an integral component in many kinds of human interaction, allowing people to act under uncertainty and with the risk of negative consequences. Privacy and trust play several roles to use any data, which are mentioned below:

- Privacy and trust is used to improve the usability of a data by a programming technique or plan or assessment.

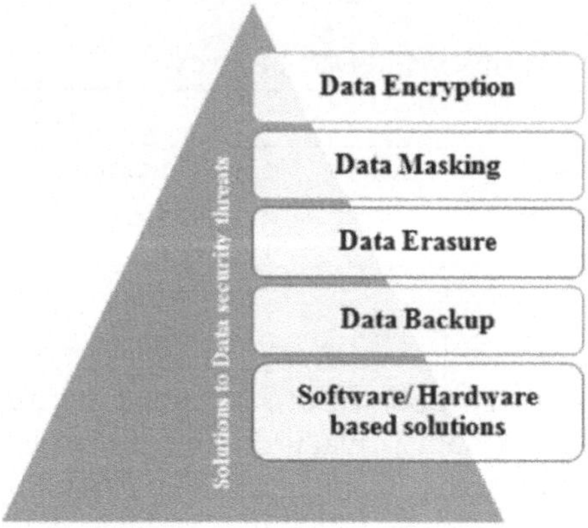

FIGURE 7.4
Solutions to data security threats.

- Governments use many privacy and trust policies to maintain the data of social networks as well as the data of the financial sectors.
- Privacy and trust can be helpful for maintaining research data.
- Privacy and trust may be useful for protecting the data communication process.

The estimation or management of trust is one of the major issues. The management of trust can be categorized into two types:

1. Lead-base frameworks
2. Notoriety frameworks

In lead-based frameworks, *trust* is considered one of the important parts for any organization. In a dispersed trust administration framework, formal tenets are utilized to express trust. The rights are conceded for the other framework in light of the standards of clients' requests for accessing the framework.

Notoriety-based administration frameworks attempt to encapsulate the rational notion of *trust*. In a notoriety framework, association and criticism are achieved by assessing members (Li et al., 2016). Positive feedback leads to improvements in notoriety. The aggregate understanding of all the members conveys the notoriety of the entire framework.

Data privacy or *information privacy* is the feature of information technology that deals with the ability of an individual or organization to determine what data in a computer system can be shared with others. The privacy of data basically refers to the appropriate use of data by authorized individuals. It basically emerges as a relationship between the storage and transmission of users' sensitive data, public anticipation of privacy and the legal issues related to it. Privacy can be synonymous with data confidentiality. However, data privacy has some features that are distinct from those of data security, as given in Table 7.1.

Protecting data privacy in an information system can be administered in two ways:

1. *Policy communication*: Privacy policies are communicated and legalized against preferences of individuals; an example is Platform for Privacy Preferences (*P3P*).

TABLE 7.1

Difference between Data Privacy and Data Security

Data Privacy	Data Security
It is about authorized access, i.e., who defines the data and who can access the data	It is about securing data against unauthorized access
It is a legal issue	It is basically a technical issue

2. *Policy enforcement*: Security policies are enforced within a security system to furnish the protection of data privacy during transmission and storage. Some examples are Extensible Access Control Markup Language (*XACML*), Enterprise Privacy Authorization Language (*EPAL*) web privacy, etc.

Nowadays, DNA computing is used for providing better security, privacy and trust (Namasudra et al., 2017c). The advantages of using DNA computing are the ability to scale capacity, store information remotely and share benefits in a dynamic domain, which can be developed into a suitable approach to maintain trust and privacy for potential users. Figure 7.5 shows the interactive representation of security, privacy and trust.

At present, a security component is required to conceal the contents of unique messages, which are directly identified with security and protection and the verification of the identity of the remote client. As a solution, Secure Public Key Infrastructure (SPKI) has been developed for chip circuits to maintain the policy in the maker. SPKI is a self-possessed infrastructure for splitting the tasks and to provide accesses of the components, which support the programmability for validating the plans and credentials. It can be considered as support validations. SPKI provides a complex foundation to conceal the message in DNA bases to ensure data security and protection (Thilagavathy and Murugan, 2016). In DNA computing, the text messages are determined by using DNA base pairs followed by the creation of an arbitrary series. In SPKI, a 4-point conversion is performed, and then, an 8-point hybrid task is performed to improve the data security. In the decryption process, SPKI performs a de-crossover and de-mutation technique by utilizing a hybrid key and a transformed key. The DNA can contain valid data that processed through polymerase chain reaction process, which are encompassed by 109 times of its size.

FIGURE 7.5
Interactive incorporation for security, privacy and trust.

7.4 Anonymity

Despite their familiarity, the concepts of privacy and anonymity are commonly misunderstood within the digital context. These concepts are foundations of the common criteria for evaluating the standard of information technology security.

Anonymity is intrinsically present in the concept of privacy. Nevertheless, anonymity refers exclusively to matters related to identity. Anonymity ensures that a user may use a resource or service without disclosing his or her identity. The main requirement of anonymity is to protect the user's identity. Anonymity is not intended to protect the subject's identity. Anonymity requires that other users or subjects are unable to determine the identity of a user connected with a subject or operation. Accordingly, the definition of anonymity is considered as the property that guarantees the user's identity from being disclosed without consent.

Anonymity can be defined as a scenario, wherein the person does not reveal his or her true identity (Tae-ho, 2015). If one stays anonymous, no one can determine who the user is or track or monitor them. In the electronic world, anonymity can be achieved, if records and transactions are anonymous, that is, if their data cannot be associated with a particular individual, either on the basis of the data itself or by combining the transaction with other data.

Anonymity is a part of our everyday life, since the activities of daily works are more or less anonymous, for example various kinds of phone or public enquiries, exchange and cash transactions.

Anonymity brings a certain kind of freedom. There are no burdens to do with background, social status, previous actions, etc. No prejudices are possible. It enables people to do things and express themselves. There are many situations in which a person might want to stay anonymous. These reasons are generally to do with psychological and social mechanisms.

A person may want to seek help or express thoughts about matters that could cause him or her social disgrace or classify them as a certain kind of person that is less socially acceptable. An example can be divergent sexual interests. Another person again may want to criticize another person or authority without getting into danger. A third person again may want to try to keep personal data out of reach of intrusive marketers and governments.

In the electronic world, there is generally less anonymity since most of the traditional processes require some kind of identification information or user authentication. In order to stay anonymous, people often have to provide false identities (*pseudonyms*), or in the worst-case scenario, they end up using someone else's identity (*identity theft*).

7.4.1 Problems with Anonymity

Since anonymous persons cannot be held accountable for their actions, anonymity attracts illegal and otherwise suspicious activities. This is seen as a major threat and is often used as an argument against anonymity. Another often-used argument against anonymity is that many interactions, for example the withdrawal of money from an account, require trust and credibility, and this requires that the person needs to be identifiable.

However, trust and credibility do not necessarily require that the user be identified. In a similar manner, the user can use different accountable pseudo-identities on different occasions, since the user may have different identities. Accountability can be guaranteed by a trusted third party. By using the information of a user, the true identity of the user

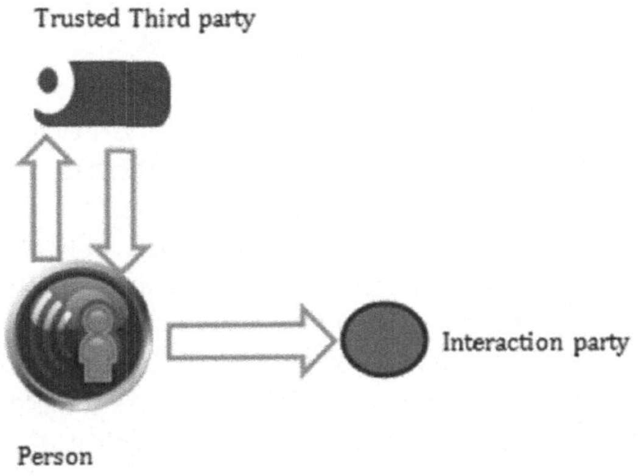

FIGURE 7.6
Accountable pseudonym.

can be linked by the trusted third party with the used pseudonym. The real identity can be determined, if there are persistent reasons for it (e.g., if there has been a crime and a search warrant is available).

Figure 7.6 depicts the concept of *accountable pseudonyms*, when a third party guarantees the credibility of the pseudonym used by the person. In Figure 7.6, the third party is able to determine the true identity of the person. Recently, several identity protection solutions have been proposed to address anonymity. Many methods advocate the use of encrypted pseudonyms or the de-identification of explicit identifiers, such as name or social security number, initially associated with genomic data. However, these solutions lack proof or guarantees of privacy afforded to the protected data. Contrary to popular belief, the protection of a patient's anonymity in genomic data is not as simple as removing or replacing explicit identifying attributes. Though genomic data may look anonymous, anonymity can only be guaranteed, when inferences can be garnered from genomic data itself.

While encryption and de-identification prevent the direct linking of genomic data to explicit identity, research shows that they provide a false appearance of anonymity. Specifically, this work is concerned with genomic data scattered across a set of locations. In a distributed data sharing environment, patients visit and leave behind data at multiple data-collecting locations, such as hospitals. Each location may separate genomic data from clinical data, and subsequently, release genomic data in order to enable such endeavors as basic research. Researchers have developed and evaluated a general technique for re-associating seemingly anonymous genomic data with the named individuals from whom the data were derived. For an example, the Online Realm is a distributed environment in which IP addresses can be re-identified with individuals. To discuss and prove the existence of a trail re-identification for a different environment, such as for health or another type of data, it must be analyzed in light of the environmental policies, oversight, methods of sharing, and data availabilities. Thus, this addresses the features that enable re-identification to occur for genomic data to serve two main purposes. First, it raises awareness of the healthcare and medical inferences that exist in a data sharing environment. Second, it provides the biomedical community with a formal computational model of a re-identification problem that pertains to genomic data.

There are several reasons for which privacy protection methods fail to sufficiently protect the anonymity of genomic data. One reason is that current methods neglect to protect identifying inferences drawn from the genomic data itself. Another reason is the ability to relate genomic information to other publicly available information.

7.4.2 Privacy and Anonymity in DNA Computing

DNA computing requires straightforward and powerful calculations. Generally, numerous researchers have developed many DNA-based encryption schemes. However, it is too soon to choose the ideal goal for some cryptographic capacities. For example, DNA verification techniques (Jung, 2010), advanced signature and secure information stockpiling (Rahman et al., 2015) are still in the initial stages (Lesk, 2007). The new frontier of direct-to-consumer genetic testing may impact on those who wish to protect their privacy and anonymity. Such testing is intrinsically and purposefully set up to identify individuals and for online information sharing. While some genealogical sites and DNA databases allow people to opt out of identification, it is possible to identify a de-identified individual from a genetic database by using statistical methods. When DNA information is combined with further metadata, such as ages, places, etc., the chances of correctly identifying the source of the DNA increase.

Currently, DNA cryptography provides a suitable solution to protect against attacks by using novel DNA-based algorithms. However, it is always problematic for the administrator to choose between an authentic data access request and a data access request from attackers or hackers. By using DNA computing, this problem can be solved.

7.5 Conclusions

Data security is a key factor in any database organization or information system. Data security is an essential feature for the effective functioning of any organization or system. In this chapter, the basic protocol of data security, threats and the various solutions are discussed in detail. Along with its underlying concepts, data security brings raises issues of privacy, trust and anonymity. The role of these sub-concepts in ensuring strong data security have been discussed in this chapter. DNA computing is a trending concept in recent years, offering high computational capability and massive storage capacity in the form of DNA bases. It offers a number of security techniques to ensure information security in a network-based data sharing environment. The applicability of DNA computing in data security and data privacy and the definition of anonymity have been discussed in detail in this chapter. The chapter offers in-depth knowledge about different aspects of data security from the perspective of DNA computing.

References

Adleman L. M. (1994). Molecular computation of solutions to combinatorial problems. *American Association of Advancement of Science*, 266(5187), 1021–1024.

Boneh, D. and Lipton, R. (1996). Making DNA computers error resistant. In *Proceedings of Second Annual Conference on DNA Based Computers*, pp. 102–110.

Data Security. (2018). Available on: https://en.wikipedia.org/wiki/Data_security.

Introduction to Information Security. (2018). Available on: https://www.cengage.com/resource_uploads/downloads/1111138214_259146.pdf.

Jung, E. J. (2010). Privacy and security definition of security/privacy attacks, services and mechanisms passive attack (1) - eavesdrop passive attack (2) - analysis.

Lesk, M. (2007). The new front line: Estonia under Cyberassault. *IEEE Security and Privacy Magazine*, 5(4), 76–79. https://doi.org/10.1109/MSP.2007.98.

Lewin, D. I. (2002). DNA computing. *Computing in Science Engineering*, 4(3), 5–8. DOI:10.1109/5992.998634.

Li, D., Dong, X. and Cao, Z. (2016). Secure and privacy-preserving pattern matching in outsourced computing. *Security and Communication Networks*, 9(16), 3444–3451.

Lipton, R. J. (1995). DNA solution of hard computational problems. *Science*, 268(5210), 542–545.

Malik, M. and Patel. T. (2016). Database security – attacks and control methods. *International Journal of Information Sciences and Techniques (IJIST)* 6(2), 175–183.

Namasudra, S. (2018). Cloud computing: A new era. *Journal of Fundamental and Applied Sciences*, 10(2), 113–135.

Namasudra, S. (2017). An improved attribute-based encryption technique towards the data security in cloud computing. *Concurrency and Computation: Practice and Exercise*. DOI:10.1002/cpe.4364.

Namasudra, S. and Roy, P. (2015). Size based access control model in cloud computing. In *Proceedings of the International Conference on Electrical Electronics, Signals, Communication and Optimization, IEEE*, Visakhapatnam, India, pp. 1–4.

Namasudra, S. and Roy, P. (2016). Secure and efficient data access control in cloud computing environment: A survey. *Multiagent and Grid Systems – An International Journal*, 12(2) 69–90.

Namasudra, S. and Roy, P. (2017a). Time saving protocol for data accessing in cloud computing. *IET Communications*, 11(10), 1558–1565.

Namasudra, S. and Roy, P. (2017b). A new secure authentication scheme for cloud computing environment. *Concurrency and Computation: Practice and Exercise*, 29(20). DOI:10.1002/cpe.3864.

Namasudra, S. and Roy, P. (2017c). A new table based protocol for data accessing in cloud computing. *Journal of Information Science and Engineering*, 33(3), 585–609.

Namasudra, S. and Roy, P. (2018). PpBAC: Popularity based access control model for cloud computing. *Journal of Organizational and End User Computing*, 30(4), 14–31.

Namasudra, S., Nath, S., and Majumder, A. (2014). Profile based access control model in cloud computing environment. In *Proceedings of the International Conference on Green Computing, Communication and Electrical Engineering, IEEE*, Coimbatore, India, pp. 1–5.

Namasudra, S., Roy, P., and Balamurugan, B. (2017a). Cloud computing: Fundamentals and research issues. In *Proceedings of the 2nd International Conference on Recent Trends and Challenges in Computational Models, IEEE*, Tindivanam, India.

Namasudra, S., Roy, P., Balamurugan, B., and Vijayakumar, P. (2017b). Data accessing based on the popularity value for cloud computing. In *Proceedings of the International Conference on Innovations in Information, Embedded and Communications Systems (ICIIECS), IEEE*, Coimbatore, India.

Namasudra, S., Roy, P., Vijayakumar, P., Audithan, S., and Balamurugan, B. (2017c). Time efficient secure DNA based access control model for cloud computing environment. *Future Generation Computer Systems*, 73, 90–105.

Nixon D. (2002). DNA and DNA computing in security practices – is the future in our genes? GSEC Assignment Version 1.3., SANS Institute.

Ogihara, M. and Ray, A. (1999). Simulating Boolean circuits on a DNA computer. *Algorithmica*, 25, 239–250.

Rahman, U., Balamurugan, N. H. C. and Mariappan, R. (2015). A novel DNA computing based encryption and decryption algorithm. *Procedia Computer Science*, 46, (ICICT 2014), 463–475. https://doi.org/10.1016/j.procs.2015.02.045.

Sarkar, S., Saha, K., Namasudra, S., and Roy, P. (2015). An efficient and time saving web service based android application. *SSRG International Journal of Computer Science and Engineering (SSRG-IJCSE)*, 2(8), 18–21.

Tae-ho, J. (2015). T control cloud data access privilege and anonymity with fully anonymous attribute-based encryption. *IEEE Transactions on Information Forensics and Security*, 10(1), 190–199.

Thilagavathy, R. and Murugan, A. (2016). Cloud computing: A survey on security issues and DNA, ID-base cryptography. *Indian Journal of Science and Technology*, 9(28). DOI:10.17485/ijst/2016/v9i28/93836.

Tornea O. and Borda M. E. (2009). DNA cryptographic algorithms. MEDITECH 2009. *IFMBE Proceedings* 26, 223–226.

WannaCryransomware attack. (2017). Available on: https://en.wikipedia.org/wiki/WannaCry_ransomware_attack

8

DNA Computing Algorithm

Awanish Kumar

CONTENTS

8.1 Introduction

Biomolecular computing has become highly important worldwide due to the massive progress that has been achieved in computational technologies using biomolecules. Among all biomolecules (DNA, RNA, and protein), researchers have extensively explored DNA as a material for computing because DNA stores all the information and instructions required for controlling a living system. It is also promising because of the incredible density of data associated with DNA. Using DNA as data, we expect and need an efficient computer to carry out the computing process. DNA computing will be harnessed to work inside the living cells/systems and combine with their existing machinery, making new methods of disease detection and possible treatment very likely. DNA is capable of storing and processing billions of times more data than existing computers; therefore, DNA computing is also known as *molecular-scale electronic technology*. DNA computing is performed with computational knowledge using robust biological molecules of DNA rather than silicon chips, as are currently used in computers. DNA computing was physically realized in 1994 when Dr. L. M. Adleman (a renowned American computer scientist) showed how DNA molecules could be used to solve biological problems (Adleman, 1994a). In DNA computing, information is represented using the four nucleotides of double-stranded DNA

(i.e., adenine [A], guanine [G], cytosine [C], and thymine [T]), where A binds with T by a double hydrogen bond and G binds with C of the opposite strand via a triple hydrogen bond. This binding is called *base-pairing* and is shown in Figure 8.1 (Watson and Crick, 1953).

One strand of DNA is known as the Watson (W) strand and other is known as Crick (C) in honor of the names of their discoverers. These two DNA single strands, with W/C complementary bases (moving in opposite orientation) at each position, bind to each other and form a double-strand DNA. DNA undergoes a process of replication and synthesizes new strands of DNA (Figure 8.2). Some algorithms are required to process the DNA data, which is defined as a step-by-step list of well-defined instructions that takes some input, processes it, and produces a meaningful result in terms of output. An algorithm's input is therefore represented by DNA molecules with specific sequences. The instructions are carried out by laboratory operations on the molecules, such as chopping DNA strands containing a certain subsequence or sorting them according to length. The result is defined as some property of the final set of molecules (i.e., the presence or absence of a specific sequence) in the DNA computing. Researchers believe that one day, molecular computers could solve the problems that would cause existing machines to struggle. This struggle would be due to the inherent massive parallelism of biological science. A small drop of water can contain trillions of DNA strands, and as biological operations act on all of them effectively in parallel (as opposed to one at a time), it was argued that one day, DNA computers could solve difficult problems that are beyond the scope of normal computers. This chapter begins with an introduction, Section 8.1, followed by a discussion of research trends in DNA computing (Section 8.2). I then continue by examining various types of DNA computing algorithms in Section 8.3. Section 8.4 is devoted to some recent algorithms of DNA computing like QPSO, VEPSO, and VEQPSO. Section 8.5 provides future research direction, and Section 8.6 of this chapter presents concluding remarks.

Since DNA computing is nascent and growing area, a lot of problems are associated with it, but a lot of hope/promise is also associated with it. DNA computers work by

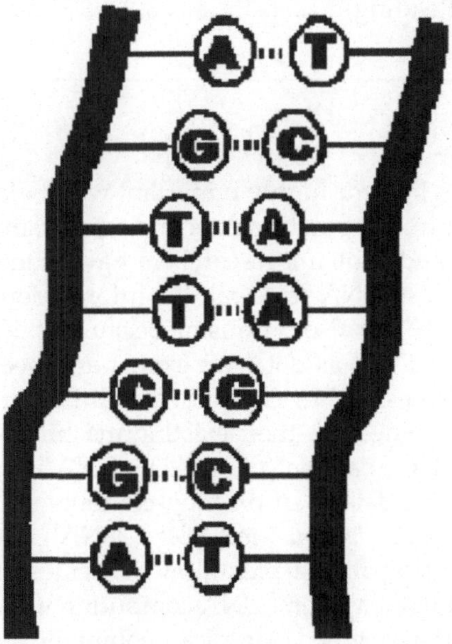

FIGURE 8.1
Double-stranded DNA molecule.

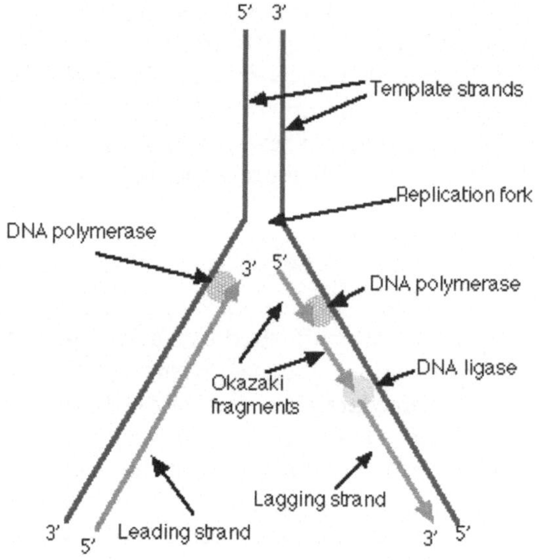

FIGURE 8.2
Scheme of DNA replication.

encoding the problem to be solved in the language of DNA (A, T, C, and G). Using this four-number system (DNA base), the solution to any possible problem can be encoded along with a DNA strand. Every possible sequence can be chemically created in a test tube on different DNA strands, and the correct sequences can be filtered out using the tools of genetic engineering. But the biggest barrier to solving large instances is that right now, we cannot synthesize very long strands of DNA arbitrarily. We can synthesize a DNA strand of a few nucleotides with no problem, but as this number increases, the yield quickly becomes too low to be practical. To represent larger problem instances, each vertex needs a unique encoding. If the encoding is too short, there will be a high probability of random sections overlapping by accident when they are not supposed to, thereby ruining the experiment. Generating solution sets even for some relatively easy problems there may require impractically large amounts of DNA. That means a huge number of DNA strands is required for the DNA computing process. Many empirical uncertainties (including those involving the generation of optimal encoding techniques, the ability to perform necessary bio-operations, and the actual error rates) are associated with DNA computing, which would be keenly addressed because computing with DNA can work in a massively parallel fashion.

8.2 Research Trends in DNA Computing

8.2.1 Pre-Adleman era

Richard Feynman was a famous American physicist who was responsible for introducing the term *molecular computation* (Heyries et al., 2009). This idea was developed by Feynman in the late 1950s. He was the first to conceive this idea and felt that an individual molecule like DNA could be used for computation for better understanding.

Therefore, the direction of innovation has been from biology to computer. One gram of DNA molecule (approximately volume 1 cm^3) can hold a lot of information in comparison to a trillion compact discs (CD; approximately 750 terabytes each) (Holland, 1992; Ulam, 1972). Primarily, the focus was on implementing aspects of living systems in computational devices. But over the next few years, nothing came to fruition or developed in the field of DNA computing. The pre-Adleman era was an unsatisfactory one for DNA computing.

8.2.2 Adleman era

The concept of DNA computing was published by Dr. Leonard Adleman in November 1994 in an article in the prestigious journal *Science* (Adleman, 1994a). He showed that DNA could be used to store data and even perform computations in a way that was massively parallel to the process used in existing computers. He was very much impressed with the idea of DNA computing when he noticed how DNA replication was remarkably similar to an early theoretical computer developed by Alan Turing in the 1930s. Researchers have applied similar ideas to solve difficult problems, like the maximal clique problem, the shortest common superstring problem, even breaking data encryption standard (DES) and 3-SAT soon after Adleman's research papers came out (Adleman, 1994a,b). So basically, the idea of DNA computing was explained and established by Dr. Adleman.

8.2.3 Post-Adleman era

There was much advancement in formalizing rules and trying to develop "universal" DNA computers from a theoretical perspective. It has been difficult to build these computers because some of the enzymes required for some operations do not yet exist. There has also been progress on the practical side. In 2006, a simplified DNA computer was built that had the ability to detect if a combination of enzymes were present and to only release medicine if they were all present (indicating that the patient had a disease). In recent years, researchers have built transcriptors (DNA versions of logic gates) to strengthen DNA computing. One reason these are important is that transcriptors are reusable, whereas previously all reagents had to be thrown away after each operation.

8.3 DNA Computing Algorithms

DNA computing is based on the idea of molecular biology processes that can be used to perform arithmetic/logic operations on information encoded as strands of DNA. Today, researchers concentrate more on developing methods for testing biochemical feasibility with wet lab experiments in the field of DNA computing algorithms. Researchers are enlarging their understanding of DNA computing by developing suitable algorithms to solve various problems. Even though developing a real DNA computer is still a long way in the future, developing a suitable algorithm is a primary step in solving problems in this direction to test and simulate the stability and reliability of DNA computing algorithms. Researchers working in this field are proposing various algorithms (Table 8.1) for DNA computing, and these are described below.

TABLE 8.1

Pros and Cons Associated with Various DNA Computing Algorithms

Algorithms	Definition	Pros	Cons
Nondeterministic algorithm	It provides different outputs for the same input on different executions.	When an exact solution is difficult or expensive to derive, then a nondeterministic algorithm is useful for finding approximate solutions.	Variations in result are produced by nondeterministic algorithm.
Adaptive algorithm	It evaluates the result obtained by each trial in the form of a concentration difference, using the detected molecule.	It is applied to the shortest path problem and the correctness is verified during the process.	An increase in the number of process repetitions may cause an explosion of calculation time.
Genetic algorithm	It offers a solution to intractable problems, that is, problems with excessive computational complexity.	It solves the intractable problem in polynomial time. It also provides a near-optimal solution.	Users may not find any satisfactory partial solutions, and tuning can also be a challenge.
Numerical algorithm	DNA serials are created in electronic computers and they are converted into numerical values.	It is fast in performing.	Data transformation is required.
Particle Swarm Optimization algorithm	It is a population-based stochastic optimization algorithm initialized with population of random solutions and searches for optima by updating generations.	PSO does not require that the optimization problem is differentiable. It is easy to implement and few parameters need to be adjusted. There are not many parameters that need to be tuned in PSO.	The performance is not competitive in some problems and the genetic operators have to be carefully selected or developed.

8.3.1 Nondeterministic Algorithm

In the deterministic algorithm, which is widely used in computer science, a particular input always produces an output. In the deterministic algorithm, the underlying machine always passes through the same sequence of states. The deterministic algorithm is the most studied and familiar kind of algorithm, and it is one of the most practical, because it can be run efficiently on real machines. But even for the same input, the nondeterministic algorithm can exhibit different behaviors on different runs. Nondeterministic algorithms are often used to find an approximation to a solution, when the exact solution would be too costly to obtain using a deterministic one. Adleman has used the nondeterministic algorithm to solve various problems. A nondeterministic algorithm is different from its more familiar deterministic counterpart in its ability to arrive at outcomes using various routes (Figure 8.3). Through nondeterministic algorithms, a large number of problems can be conceptualized in DNA computing theory. This type of algorithm is used to solve problems that allow for multiple outcomes.

8.3.2 Adaptive DNA-Based Computing Algorithm

This is a method that is based on the concept of a popular molecular biology technique called polymerase chain reaction (PCR). It could improve the explosion problem of the

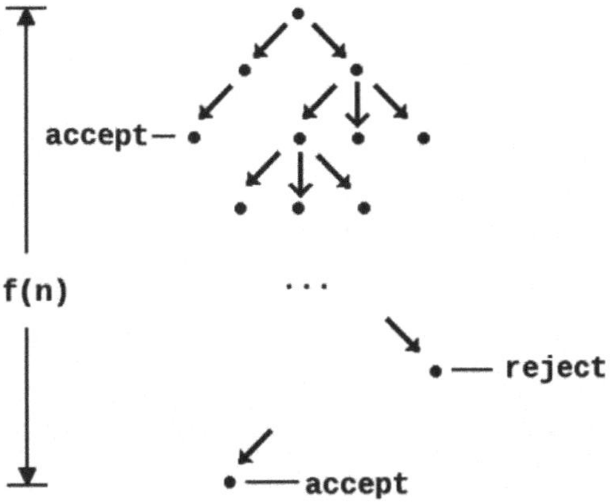

FIGURE 8.3
Nondeterministic algorithm exhibits different behaviors on different runs, even for the same input.

DNA molecules. The DNA-based adaptive algorithm consists of various steps: (1) the given problem is encoded by the molecules of DNA, (2) the coded DNA molecules are made to connect, (3) the candidate is extracted from the generated combinations and this combination could be a solution, (4) the extracted combination is incised by using an enzyme restriction endonuclease, (5) the fragmented combination is amplified by using the PCR technique, and (6) finally steps 2–5 are repeated again (Watanabe et al., 2004). The adaptive algorithm is designed on the basis of Adleman-Lipton paradigm of DNA computing. However, unlike the Adleman-Lipton architecture, a cutting operation has been introduced to the algorithm and a mechanism by which the molecules used by computation were fed back to the next cycle was devised. The proposed adaptive algorithm (PCR based) is applied to the shortest path problem, and the correctness is verified by performing a computational simulation. The adaptive algorithm adjusts the concentration value of the molecules by one cycle computation and reuses the results of concentration adjustment in the next cycles. Therefore, this algorithm does not need to search for a solution in a series of operations. Thus, by repeating the cycle of connection (i.e., extraction, cutting, and amplification), the generation probability of combinations serves as a solution. The candidate sequence is raised and finally, the adaptive algorithm detects an optimal solution. It is verified by actually performing a simulation, although it is important how many times the repeated calculation is performed. The adaptive technique evaluates the result obtained by each trial in the form of a concentration difference and finally uses the detected molecule. However, there is a problem with adaptive algorithms. There may be a limited repetition of chemistry operation, and the increase in the number of repetitions may cause an explosion of the calculation time (Watanabe et al., 2004).

8.3.3 Genetic Algorithm (GA)

This is a method for solving both unconstrained and constrained optimization problems based on a natural selection process that mimics biological evolution. The algorithm repeatedly modifies a population of individual solutions. The GA randomly selects individuals from the current population at each step and uses them as parents

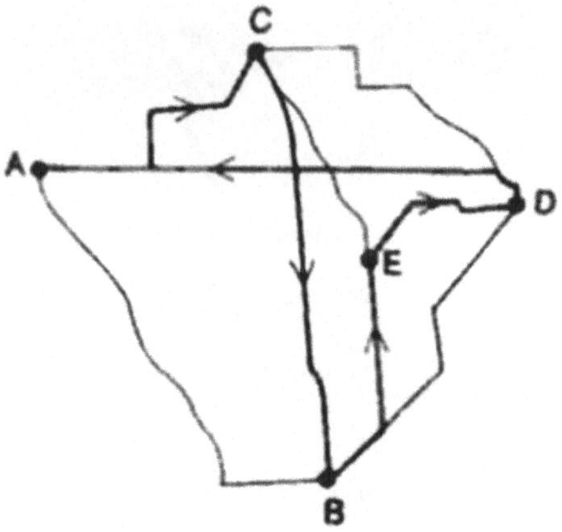

FIGURE 8.4
Result after crossover operations in GA.

to produce the children for the next generation. The population "evolves" toward an optimal solution over successive generations (Yuan et al., 2004). GA can be applied to solve problems that are not well suited to standard optimization algorithms, including problems in which the objective function is discontinuous, stochastic, nondifferentiable, or highly nonlinear. GA generates a population of points at each repetition for the optimal solution of the process. It selects the next population by a computation that uses random-number generators. GA uses three main types of rules at each step to create the next generation from the current population: (1) Selection rules: It selects the individuals, called *parents*, that contribute to the population at the next generation; (2) Crossover rules: It combines two parents to form children for the next generation; and (3) Mutation rules: It applies random changes to individual parents to form children. GA performs crossover operations nicely. Basic crossovers might result in invalid members of the population. For example, please label few places on Earth with A, B, C, D, E, F, or other notation. After combining the points ABCDEF and BDAFCE, that is, the crossover operation, GA gives result into ABCFCE (Figure 8.4). It is a parallel, global optimization method with the search strategy and supportive for DNA computing. GA may be one of the possible ways to break the barrier of DNA computing and to make it practical as the problem size scales up.

8.3.4 Numerical DNA Computing Algorithm

In this algorithm, DNA serials are generated in electronic computers and they are converted into numerical values for computing (Figure 8.5). Although the cost of computing performed electronically is less, the mentioned advantages of DNA molecules cannot be applied to problems entirely because when the DNA molecules are used in the solution environment, they only can use the abovementioned characteristics. It uses A, T, G, and C bases rather than a dual number of systems, and the solution sets of the problems are composed of these bases (Xu et al., 2011). Conversion of DNA serials into numerical data are shown in Figure 8.5. Coding can be created randomly using A, G, C, and T bases and converted into numerical data using 0, 1, 2, and 3 values, respectively.

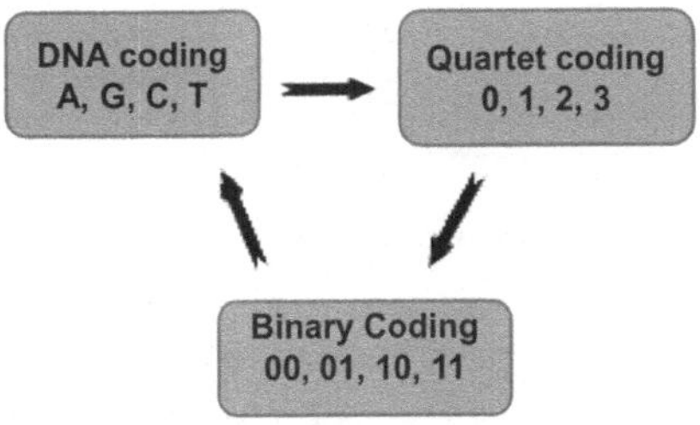

FIGURE 8.5
Conversion of DNA serials into numerical data.

8.3.5 Particle Swarm Optimization (PSO)

This is widely used for solving various optimization problems. This algorithm, with its stochastic means, is well equipped to handle a number of problems. PSO was introduced by Eberhart and Kennedy in 1995, inspired by the social behavior of animals such as fish schooling, bird flocking, and swarm theory. PSO has some attractive characteristics, and it has proven to be more effective in many cases than GA and other similar evolutionary techniques (Hassan et al., 2005). PSO has been effectively applied in many fields like artificial neural network training, function optimization, fuzzy system control, and other areas where GA can be applied. GA and PSO have both been used extensively for a variety of optimization problems, and in most cases, PSO has proven to have superior computational efficiency. In comparison to GA, PSO is more advantageous and easy to implement. Many attempts have been made to improve the performance of the PSO since 1995, and a quantum theory has been introduced into the PSO. Researchers proposed it as a quantum-behaved PSO (QPSO) algorithm, which is guaranteed theoretically to find good optimal solutions in search space and to find out or evaluate the particle position (Figure 8.6).

The experimental results on some widely used benchmark functions show that the QPSO works better than standard PSO. Various algorithms that are based on the PSO principles are discussed in the next segment (Section 8.4).

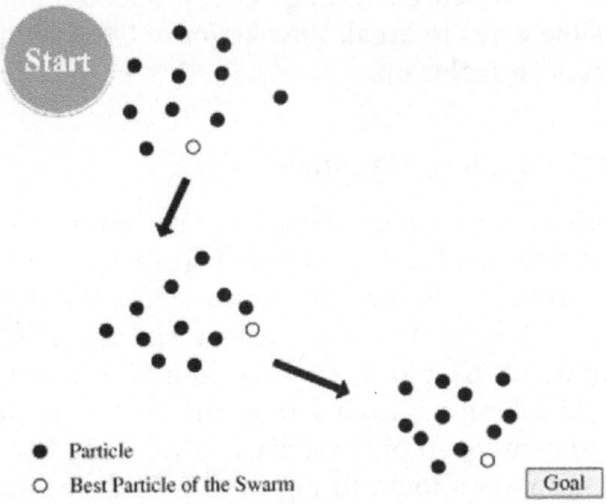

FIGURE 8.6
PSO used to evaluate the particle position and update the location of each particle in every iteration.

8.4 Quantum-Behaved Particle Swarm Optimization (QPSO)–Based Algorithms

QPSO is based on an adaptive DNA computing algorithm. It is supported and based on various parameters and attributes: (1) Parameters of population size, rate of enzyme mutation, maximum number of operations, virus mutation rate, crossover rate, and fitness function of DNA computing algorithm are concurrently tuned for the adaptive process; (2) the numerical realization of DNA computing algorithms with the proposed approach is implemented in the system identification; and (3) the adaptive algorithm is performed using QPSO algorithm for faster operation, goal-driven progress, and flexibility in the data. QPSO has a great capacity for global searching. This algorithm has been inspired by living beings that move as a mass, including insects and bees, and it was used for the solution of many problems in the literature (Karakose and Cigdem, 2013). The QPSO algorithm is a population-based algorithm, in which individuals act together and make up the masses. The next position of the population is determined by taking the average of the best results acquired by the particles that make up the population in QPSO in the vicinity, while the speed of the population is used in this algorithm (Xi et al., 2007).

Capability and effectiveness of different DNA computing algorithms vary. The size of the population suitable for QPSO, enzyme proportion, maximum operation number, crossover rate, and virus proportion affect the local and global search capacity of the algorithm. The real target, aimed at adapting the parameters, is to increase the diversity of the population and prevent a focus on local optimum points. Furthermore, adaptable algorithms can easily be applied to the solution of many optimization problems. The abovementioned parameters, which are fixed at the beginning, make it dynamic and increase the effectiveness of the QPSO algorithm. A new QPSO-based adaptive DNA computing algorithm is proposed and implemented for optimization problems in different applications. QPSO approach has high accuracy, effective optimization, and high convergence speed as compared to traditional DNA computing algorithms. The general search capability of the method is not dependent on local optimum points. The applicability of the QPSO approach is easier, simpler, and faster than traditional DNA computing algorithms (Karakose and Cigdem, 2013). DNA computing algorithms have some limitations in terms of adaptability, convergence speed, and effectiveness. QPSO is a good approach that aims to perform DNA computing algorithms with adaptive parameters.

Vector Evaluated Particle Swarm Optimization (VEPSO) is a new approach to improving DNA computing that is currently being proposed and advocated. The generic composite design optimizes the framework being presented in the current work, employs a vector evaluated technique for multiobjective optimization like VEPSO, and was used by Omkar et al. (2008). The vector-evaluated QPSO (abbreviated as VEQPSO) allows the evaluation of the multiple objectives to be separated, which proves to be very appropriate for the current problem. This is a co-evolutionary method that employs separate swarms for each of the objective and information migration between these swarms, ensuring an optimal solution with respect to all objectives. The VEQPSO is a multiobjective QPSO method inspired by the concept and ideas of VEGA (Schaffer, 1984) and VEPSO algorithm (Omkar et al., 2008). The VEQPSO algorithm is conceptually simple. It is similar to two single objective functions being separately evaluated by separate swarms. The multiple objectives being considered here are disparate in nature, and hence, separate and exclusive evaluation of the multiple objectives is more appropriate. Using co-evolutionary

techniques, VEQPSO is very well suited to the current challenge, as it is capable of searching for multiple optimal solutions in a very vast solution space in a single run. The key issue in these co-evolutionary algorithms is that the fitness of an individual in a population depends on the individuals of a different population. This enhances the capability of the algorithm to better explore and exploit its features.

8.5 Upcoming Research Direction in DNA Computing

Even though a lot of problems are associated with DNA computing, it has a very high capability and future potential because of its efficiency and speed. DNA molecules can store a huge volume of information in comparison to any existing computer memory chip. This means that DNA computing offers the possibility of a far denser packing of molecular information than silicon-based computers can achieve. A single bacterial cell measures just a micron square, which is about the same size as a single silicon transistor, but it holds more than a megabyte of DNA memory. It has all the computational structures to sense and respond to its environment. It has been estimated that a gram of DNA can hold as much information as a trillion CDs (to try to put this in some understandable perspective). So, DNA molecules could have mega-memory as compared to silicon chips. Hundreds of trillions of DNA molecules can be operated in parallel in a biochemical reaction. DNA computers could store a bit, 0 or 1, of data in one cubic nanometer, one trillionth the size of a conventional computer's electronic storage. Thus, a DNA computer could store massive quantities of information in the space that a standard existing computer would use to store much less. It would be about twice as fast as the supercomputer (the fastest computer of this age), performing numerous instructions per second at a time. Due to its enormous potential, future research should be emphasized to scale up DNA computing.

The current status of DNA computing research does not suggest that DNA computers will provide a successor to silicon within the next few decades. We do not believe that DNA computing should be written off completely, however, although a DNA computer in the traditional sense may be a flight of fancy. There are important application areas where the technology may play a part, and it may be possible that DNA computing technology can be integrated with more traditional approaches to create DNA/silicon hybrid architectures or within software. Since the software is more flexible and suited to rapid adaptation than hardware, we may see DNA computing benefits being implemented and exploited by the software first, leaving hardware to play catch-up. Success pivots on the improvement of the DNA computing process to reduce the time taken to isolate the correct results from all the possibilities created. The addition of autonomy is required to allow DNA computers to arrive at their results with the minimum of human interference.

DNA computing continues to be an exciting concept and one delicately placed at the edge of biology and science. Not only it has become a promising technology for analyzing data, but it also shows power and ability to transmit information in nanotechnology and other interesting advantages and applications (Figure 8.7). DNA computing will hopefully overcome its current challenges with continued research and development and will pave the way for successful and efficient computing application in a wide variety of fields. We believe that within a few years, we will see scaling up by a factor of a trillion or more because the biochemical operations involved are often slow and subject to error. There

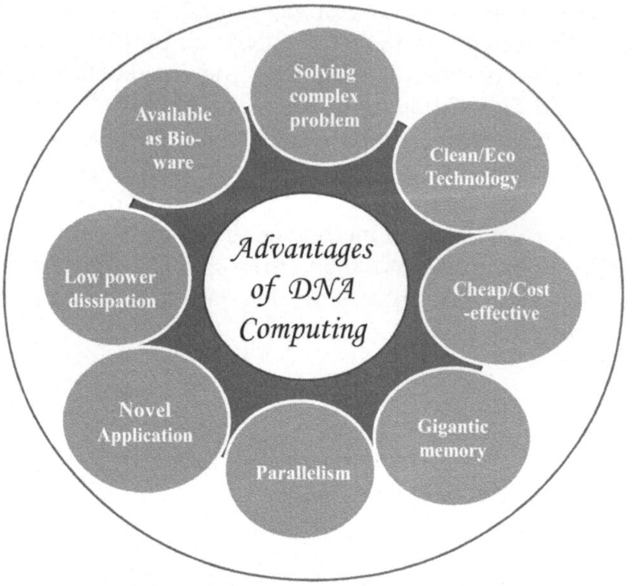

FIGURE 8.7
Advantages and applications of DNA computing.

are still some obstacles in employing DNA computation to its full potential, and rigorous tests of its accuracy and further technological developments are needed to achieve regular practice and implementation.

8.6 Conclusions and Future Work

The ability to program a living cell or system not only offers interesting possibilities in computational science but also develops a new tool for use in the medical and biological sciences. If researchers could accurately utilize the genetic mechanisms of a cell for accomplishing manmade tasks, it is theoretically possible that a collection of such cells could be used as biological nanomachines to carry out many of desirable medical functions. The field of DNA computing is in its infancy. While DNA computing has promise, new challenges are emerging every day. The most important challenge is uncertainty, because of DNA chemistry, computational results, and the exponential increase in the number of DNA molecules necessary to solve problems of interesting size. Despite these issues, specific progress has been made both in the development of new protocols for quantifying errors and in creating more efficient and error-tolerant DNA computing. In addition to this, new paradigms based on molecular evolution have emerged from molecular biology to initiate a new direction in DNA computing. Only further work will allow the determination of the proper scope of and niche for DNA computing. In conclusion, DNA computing is one of the youngest, most amazing, and most exciting areas to be explored or developed by researchers in recent times.

As has been the case in the recent development of DNA computing, future directional and dedicated work is required to establish it. DNA computers can do 330 trillion operations per second, which is more than 100,000 times the speed of the present fastest silicon-based computer (Xue and Xiyu, 2013). In future, we may expect to see a hybrid device that uses traditional silicon for normal computing tasks and DNA co-processors for specific and quick functions. As new challenges emerge, the field of DNA computing remains promising and alive, and the continued exploration of current protocols and techniques may eventually result in a new body of methods that are specifically adapted to successful computing with DNA. Substantial and keen effort should be generated in the direction of data memory storage elements, processing elements, and communication elements to produce a DNA computer.

References

Adleman, L. M. (1994a). Molecular computation of solutions to combinatorial problems. *Science*, 266, 1021–1024.

Adleman, L. M. (1994b). Molecular computation of solutions to combinatorial problems, *Nature*, 369, 40.

Eberhart, R. C. and Kennedy, J. A. (1995). New optimizer using particles swarm theory. In *Sixth International Symposium on Micro Machine and Human Science*, Nagoya, Japan, pp. 39–43.

Hassan, R., Cohanim, B. K., Weck, O. D., and Venter, G. A. (2005). Comparison of particle swarm optimization and the genetic algorithm. In *Proceedings of the 46th AIAA/ASME/ASCE/AHS/ ASC Structures, Structural Dynamics and Materials Conference*, Austin, TX.

Heyries, K. A., Blum L. J., and Marquette, C. A. (2009). Straightforward protein immobilization using redox-initiated poly(methyl methacrylate) polymerization. *Langmuir*, 25, 661–614.

Holland J. H. (1992). *Adaptation in Natural and Artificial Systems*. Cambridge, MA: MIT Press.

Karakose M. and Cigdem U. (2013). QPSO-based adaptive DNA computing algorithm. *The Scientific World Journal*, Article ID 160687, 1–8.

Lipton R. (1995). DNA solution of hard computational problems. *Science*, 268, 1995.

Omkar S. N., Mudigere D., Naik G. N., and Gopalakrishnan S., (2008). Vector evaluated particle swarm optimization (VEPSO) for multi-objective design optimization of composite structures. *Computers and Structures*, 86(1–2), 1–14.

Schaffer, J. D. (1984). Multi-objective optimization with vector evaluated genetic algorithms. Ph.D. Thesis, Vanderbilt University: Nashville, TN.

Ulam, S. (1972). On some mathematical problems connected with patterns of growth of gures. In Burks, A. (Ed.), *Essays on Cellular Automata*, pp. 219–231.

Watson, J. D. and Crick, F. H. C. (1953). Molecular structure of nucleic acids: A structure for deoxyribose nucleic acid. *Nature*, 171, 737–738.

Watanabe, S., Tsuboi, Y., Ibrahim, Z., and Ono, O. (2004). Adaptive DNA computing algorithm by using PCR and restriction enzyme. *IEEE Xplore, Conference: Cybernetics and Intelligent Systems, 2004 IEEE Conference on Volume: 1.*

Xi, M., Sun, J., and Xu, W. (2007). Parameter optimization of PID controller based on Quantum-Behaved Particle Swarm Optimization Algorithm. *Complex Systems and Applications-Modelling*, 14, supplement 2, pp. 603–607.

Xu, J., Qiang, X., Yang, Y. et al. (2011). An unenumerative DNA computing model for vertex coloring problem. *IEEE Transactions on NanoBioscience*, 10(2), pp. 94–98.

Xue, B. and Xiyu, L. (2013). Design of E-commerce development roadmap for SMEs using DNA computing technique. *IEEE 17th International Conference on Computer Supported Cooperative Work in Design (CSCWD).*

Yuan, L., Fang, C. and Qi, Y. (2004). Genetic algorithm in DNA computing: A solution to the maximal clique problem. *Chinese Science Bulletin*, 49 (9) 967–971.

Further Reading

Kon, Y., Yabe, K., Rajaee, N., and Ono, O. (2007). Adaptive DNA computing algorithm by using PCR and restriction enzyme. In Park, J. W., Kim, T. G., and Kim, Y. B. (Eds). AsiaSim 2007. *Communications in Computer and Information Science*, vol 5. Berlin, Heidelberg: Springer.

References (for Advanced Reading)

Ileri, A. M., Ozercan, H. I., Gundogdu, A., Senol, A. K., Ozkaya, M. Y., and Alkan, C. (2016). Coinami: A cryptocurrency with DNA sequence alignment as proof-of-work. arXiv preprint arXiv:1602.03031.

Limaye, R., Kumar, L., and Limaye, N. (2018). Fourth generation technologies in pharmaceuticals – Revolutionizing healthcare. *Journal of Systems Biology and Proteome Research*, 2 (1): 3–7.

Schulte, P. (2017). Mobile technology: The new banking model connecting lending to the social network. In *Handbook of Blockchain, Digital Finance, and Inclusion*, 2 (13), 331–359.

Swan, M. (2015). Blockchain: Blueprint for a new economy. Sebastopol, CA: O'Reilly Media, Inc.

9

Applications and Future Trends of DNA Computing

Suyel Namasudra, Ganesh Chandra Deka, and Rohan Bali

CONTENTS

9.1 Introduction

Deoxyribonucleic acid or DNA is a lengthy molecule that holds most of the genetic details used for reproduction and growth (Wikipedia, 2018). In the human body, all the cells have the same DNA, and DNA is responsible for all the aspects of the cell. There are two main biopolymer strands in DNA molecule, which are curved around each other to form a double helix. *Nucleotides* are the fundamental block of each DNA, and they are made by nucleic acids. Each nucleotide has three units:

1. Sugar molecule
2. Phosphate group
3. Nitrogen base

There are four types of nitrogen bases:

1. Adenine (A)
2. Guanine (G)
3. Cytosine (C)
4. Thymine (T)

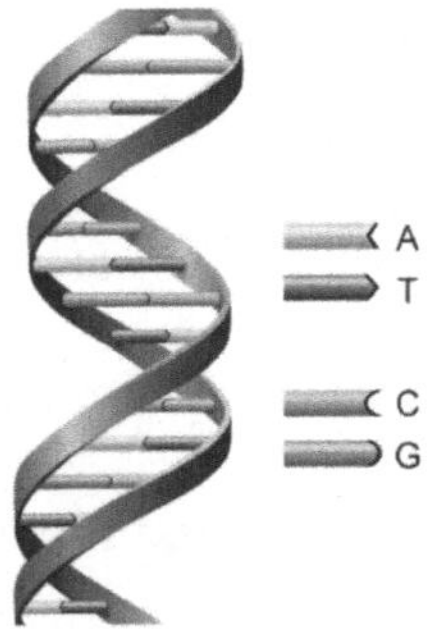

FIGURE 9.1
Structure of DNA.

"A" is paired with "T" and "C" is paired with "G" according to the rules of base pairing. A huge number of DNA sequences can be generated by using the combinations of random DNA bases. Figure 9.1 shows the structure of DNA.

DNA computing is one of the computing areas in which DNA, molecular biology, hardware and biochemistry are utilized to encode genetic information in digital computers. In 1994, Adleman (1994) was the first to use DNA in the field of computation. Adleman solved the seven-point "Hamiltonian Path Problem" by using DNA. Since then, many novel techniques have been designed using DNA computing to solve critical problems, such as the SAT problem, the 0-1 planning problem, the integer planning problem, graph theory, the optimal problem, databases, cryptography, etc., and many Turing machines have been proved by using DNA computing. DNA computing is popular for three main reasons:

1. Speed
2. Minimal storage requirements
3. Minimal power requirements.

Combinations of DNA strands have a computation power equivalent to 10^9 or more, and DNA can store memory space at the density of approximately 1 bit per cubic nanometer (Somnath et al., 2010). This leads many researchers to use DNA computing in many fields.

Recently, DNA computing has become very popular in various fields, such as security, quantum computing, etc. Because of all the data being transmitted over the internet and because of the presence of many attackers and malicious users on the internet, data security has become very important. *DNA cryptography* plays an important role in the secure communication of users' confidential and sensitive data over the network. With its rapid growth, DNA computing is also used in the field of quantum computing. In quantum computing, computers are developed that support scientific theory. Researchers are also trying to incorporate the concept of DNA computing into quantum computing. In the cloud environment, DNA computing has a significant impact on improving security because there are numerous attackers and malicious users in the cloud environment (Deka and Das, 2014; Namasudra et al., 2014, 2017, 2017a; Changxiang et al., 2007; Namasudra, 2018). Many researchers have proposed many schemes using DNA computing for the cloud environment (Sureshraj and Bhaskaran, 2012).

In this chapter, all the applications of DNA computing in various fields are discussed in detail. Moreover, future trends in DNA computing, such as cognitive computing, quantum computing, DNA sequence marketing, etc. are also presented, which can be highly beneficial for research in the field of DNA computing.

The rest of the chapter is structured into several parts. Section 9.2 discusses the applications of DNA computing. Future trends of DNA computing are presented in Section 9.3. Finally, Section 9.4 concludes the whole chapter.

9.2 Applications of DNA Computing

Nowadays, DNA computing is used in many fields. In this section, many applications of DNA computing are discussed.

9.2.1 DNA Cryptography

Cryptography is one of the major applications of DNA computing. In cryptography, symmetric and asymmetric techniques are used to convert the plaintext of any data into its ciphertext and vice versa (Terec et al., 2011). In DNA cryptography, DNA computing is used for data encryption. Many researchers have proposed many data security schemes based on DNA computing, which can be classified into two types, namely DNA-based data hiding schemes and DNA-based encryption schemes. In DNA-based encryption schemes, data are saved in the form of DNA bases (A, T, G and C) rather than in the form of 0 and 1, as in traditional cryptography approaches. Basically, the plaintext of the data is saved in the form of a DNA sequence. Then, complementary pair rules are applied to the DNA sequence to make it complex and unpredictable. Sometimes, primers and codons are also used to make the DNA sequence more complex. Thus, DNA-based encryption schemes improve data security against hackers and malicious users. In DNA-based data hiding schemes, data are hidden in the DNA sequence. Here, the existence of the data is hidden by using DNA computing to protect it against attackers and malicious users, which is known as *DNA steganography*. In this technique, the sender of any data first converts it into binary form, and then, this binary form of data is converted into a DNA sequence by using any binary encoding rule. Finally, this DNA sequence in mixed with other fake DNA sequences, and the fake DNA sequence is sent to the respective authorized user. From the fake DNA sequence, the receiver recovers the original data content. However, DNA cryptography is still in the early stages, and many researchers are trying to develop complex cryptosystems using DNA computing. Nowadays, to improve the security of the cloud computing environment and wireless sensor network, DNA computing is being used.

9.2.1.1 Data Security of the Cloud Computing Environment

Cloud computing can be considered as one of the most promising technologies in the field of distributed computing. There are mainly three entities in cloud computing, namely the cloud service provider, the data owner and the user (Namasudra and Roy, 2015, 2016, 2017a,b,c, 2018; Namasudra et al., 2017c). The cloud service provider provides many services and infrastructure to both the users and data owners. Many organizations provide cloud services, such as Google, IBM, etc. (Sarkar et al., 2015). However, security is one of the major factors in the cloud environment because of its internet-based services, and many customers and users do not want to use cloud services because of the security issue. Nowadays, DNA cryptography is widely used in the cloud environment to improve data security. Many researchers have proposed many schemes to improve data security, such as

DNA Encryption Algorithms (DNAEA) (Javheri and Kulkarni, 2014). One encryption and decryption technique of the cloud environment is discussed below.

Encryption: There are several steps in the encryption phase:

Step 1: In the first step, the data owner converts the plaintext of the data into binary form.

Step 2: The binary form of the plaintext is then converted into the DNA sequence by using any binary coding rule.

Step 3: In the third step, the data owner selects a secret key and converts it into the DNA sequence.

Step 4: In the fourth step, the data owner makes the secret key the same length as the plaintext. Then, the Exclusive OR (XOR) operation is performed between the plaintext of the data and the secret key.

Step 5: The data owner stores the encrypted data on the cloud server and sends the secret key and binary coding rule to the respective user.

Decryption: Since the respective authorized user has both the binary coding rule and the secret key, s/he can easily access the original data.

Step 1: The data owner downloads the encrypted data from the cloud server.

Step 2: In the second step, the user performs the XOR operation between the encrypted data and the secret key.

Step 3: Finally, the user performs the binary coding rule to get the original data content.

9.2.1.2 Data Security of the Wireless Sensor Network (WSN)

The *wireless sensor network* is basically a wireless network consisting of many distributed autonomous sensors used to monitor the environment. The WSN measures and monitors conditions of the environment, such as temperature, humidity, wind, sound, pollution levels and so on. It is similar to the *wireless ad hoc network* since it depends on wireless connectivity, and data are continuously transported wirelessly. However, data security is a major concern of the WSN because of the presence of many attackers and malicious users. Many researchers are now trying to improve the security of the WSN by using DNA computing. Here, DNA cryptography is used along with the Secure Socket Layer (SSL) protocol, which supports a secure channel or medium to exchange information over the WSN (Monika and Upadhyaya, 2015). There are two main phases in improving the security of the WSN:

1. *Encryption*: The encryption phase involves the following steps:

Step 1: In the first step, the public key is exchanged between two sensor nodes by using the SSL protocol.

Step 2: There are several processes in the second step to improve data security:

a. First, the plaintext of the data is converted into American Standard Code for Information Interchange (ASCII) values.
b. Then, the ASCII values of the data are encrypted by using the recipient's public key.

TABLE 9.1
Binary Coding Rule

DNA Bases	Binary Form
A	00
C	01
G	10
T	11

c. The result of the above sub-step is divided into three groups.
d. Then, each group is converted into the binary value.

Step 3: Finally, the binary values are converted into the DNA sequence by using Table 9.1. Figure 9.2 shows the flowchart of the encryption process.

2. *Decryption*: The ecryption process is simply the reverse process of the encryption phase. Since the recipient has all the information, s/he can easily decrypt the data.

Here, data are concealed by using the DNA bases, namely A, C, T and G, which enables three-layered security for the WSN.

9.2.2 DNA Fountain Strategy

The *DNA fountain strategy* includes the storage and retrieval of DNA information as Sudoku puzzles. This strategy can store ten times more data (IDC, 2014). The success of the DNA fountain strategy is mainly lies in the coding theory of the domain-specific constraints of

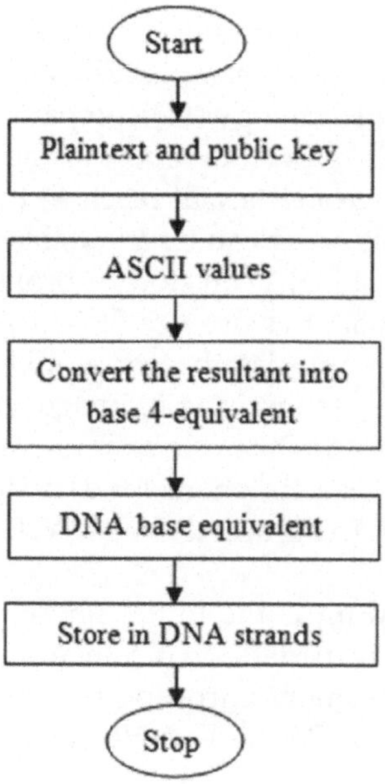

FIGURE 9.2
Encryption process for WSN.

DNA-based data storage. When this strategy is applied on DNA storage devices, it uses maximum storage capacity, while providing robustness against data corruption.

In the new study, it has been proved that streaming video on a cellular or mobile phone based on DNA computing can utilize minimum storage space, and this technique is reliable. There are two main phases in this strategy:

1. *Initial key generation*: In initial key generation, an explicit value is taken as input, and it generates a new secret key. This secret key consists of a random range followed by a group of numbers, and
2. *Ciphertext generation*: There are several steps in ciphertext generation:

Step 1: In the first step of DNA ciphertext generation, data of 128 bits in size is taken and the same length of key is selected, which is basically a DNA sequence.

Step 2: The original data is regenerated by using ASCII values, and it is stored in the 4×4 information matrix form. The 4×4 matrix includes a sequence of 0, 1, 2 and 3 for each respective row.

Step 3: As per the Advanced Encryption Standard (AES) algorithm, the matrix manipulation operation is performed in cycles, and the first row of the matrix is shifted.

Step 4: In the fourth step, for every cycle manipulation, the XOR operation is performed between the result obtained in the previous step and the key. If n is the length of the initial key, the number of the cycle iteration is $2n$.

The DNA-based secret key always makes the plaintext different from the ciphertext. Therefore, this feature supports the security of the users' confidential or sensitive data.

9.2.3 Big Data Storage

Big Data can be considered as huge amounts of data that cannot be processed by traditional data-processing techniques (Namasudra et al., 2017b). With the huge expansion of Big Data in each and every Information Technology (IT) company, security and storage issues are increasing (Hakami et al., 2015). As per the report of the International Data Corporation Firm of Framingham, the amount of data will reach 44 trillion GB by 2020, which will be very difficult to store. DNA computing can be a suitable solution for handling this huge amount of data, since one gram of DNA can store approximately 700 TB of data. So, a few grams of DNA can store all the data in existence.

DNACloud is a tool for storing any data based on a DNA sequence (Shah et al., 2013). It converts any type of data, such as text, audio, image, video, etc., into a DNA sequence, and enables it to be stored on DNA. Data can be retrieved on demand, when it is stored on DNA. DNACloud also provides estimations for data storage along with the encryption and decryption facility, as shown in Figure 9.3. In DNACloud, there are three main phases:

1. *DNA encoding*: DNA encoding is the initial phase in which a perfect technique is chosen to encode the given data or message into the DNA sequence. Many researchers have proposed many encoding techniques for converting any data into DNA sequences (Arita, 2004). The Huffman code is one of the efficient

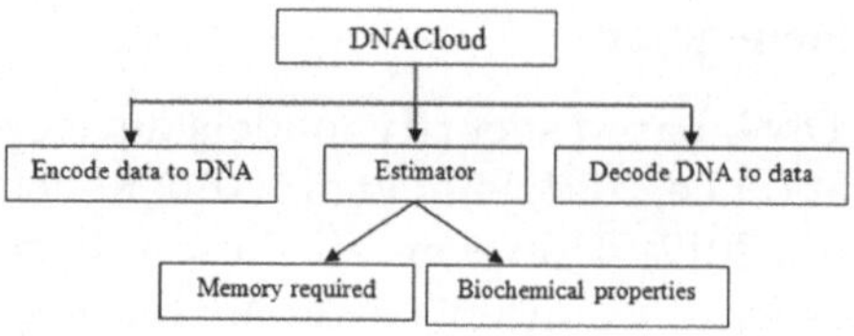

FIGURE 9.3
Functionality of DNACloud.

coding techniques, which is mainly known for data compression (Huffman, 1952). Here, DNA encoding is uniquely decodable. In DNACloud, a technique similar to Huffman encoding has been implemented (Goldman, 2013). The overlapping codes have been implemented in DNACloud for error correction. Here, the encoding module receives the data of any format as an input. The encoded DNA sequence is divided into chunks of DNA, and then, each of those DNA chunks is overlapped for error correction. Then, the original file or data is transformed into the Huffman base 3 code, consisting of a code length of 5; this is converted to triplet codon to DNA code as per the conversion principle. If any DNA base or DNA chunk is deleted, it can only be regenerated by reading the overlapped sequence code, which improves the efficiency of DNACloud.

2. *DNA decoding*: To retrieve any data stored on DNA, the reverse of the processes of the DNA encoding phase are executed. Here, data are retrieved without including the index bits, and Huffman DNA base 3 codes are converted into the original data. In this module, the DNA sequence is taken as the input and output of the original data.
3. *Storage estimator*: This module provides biochemical properties and various statistics for the encoded file. For storing data on DNA, these values of estimation help for the experiments. There are two main factors in this storage estimator:
   i. *Memory required*: Here, users can select or choose the data or file that needs to be encrypted from the system. It helps users to monitor the memory taken to encode and store any data in DNA. The following details of any file or data are estimated:
      a. Size of file
      b. Free memory required
      c. Size of the DNA string
      d. Amount of DNA required
   ii. *Biochemical properties and cost*: Here, the biochemical properties of any DNA sequence are estimated to store the data, which allows the user to devise a budget for the experiments.

9.3 Future Trends

DNA computing can be used in many fields in the future. In this section, future trends in DNA computing are discussed in detail.

9.3.1 Cryptography and Stenography

There are many existing DNA-based security models developed by many researchers (Tanaka et al., 2005; Hsu and Lee, 2006; Shiu et al., 2010; Kencla and Loebl, 2010; Zhang and Gao, 2015; Wang et al., 2017). However, all these schemes have some disadvantages. Most of the schemes mainly concentrate on increasing the complexity by introducing some complex operations. Thus, these schemes increase the time needed for data encryption. Automatically, the data accessing time is also increased. This issue becomes a major concern for the cloud computing environment, where users pay money on the basis of pay-per-use (Abbasy and Shanmugam, 2011). Because the data accessing time is increased, they have to pay more for using more cloud services. So, a novel DNA-based security model can be developed that takes a less time for data encryption.

In DNA steganography, the existence of data is hidden in fake DNA sequences to protect the data against hackers and unauthorized users (Johnson and Jajodia, 1998). A long fake DNA sequence is taken for this process. Sometimes, it is very difficult to retrieve the data from the fake long DNA sequence. There is also a chance that the data will be lost at the time of data retrieval. Thus, users' confidential or personal data can face security issues. A novel DNA-based steganography technique can be developed to protect the user's confidential data.

In both DNA-based data encryption and DNA-based data hiding schemes, DNA reference key, primers, codons, etc. are exchanged between the sender and receiver prior to the data communication process, which is one of the major issues of both these techniques. If any hacker, attacker or malicious user gets this information in the middle of the data communication process, users' confidential data may face security issues. Nowadays, most data communication processes are executed over the internet, which makes the situation more critical. Novel DNA-based security models can be developed that can overcome the above-mentioned issues to improve the data security of the user's confidential data or file.

9.3.2 Cognitive Machines

Cognitive computing can be defined as the simulation process of human thinking in a computerized model. The main goal of cognitive computing is to develop an automated digital computer system that can solve difficult problems without human assistance.

Nowadays, researchers in South Korea are trying to develop cognitive machines that are based on the convergence of biological and physical intelligence, and also trying to design biologically inspired robots based on DNA computing (Chandra, 2017). This machine can encode information based on DNA computing. Here, single-strand and double-strand DNA will be used for encoding the data and also even sometimes for software. Researchers are demanding that these systems will be able to compute data mining techniques, including dynamic programming, neural networks, decision trees, rule-based methods, probability reasoning, genetic programming, hidden Markov chains, etc. Here, a perfect DNA sequence must be selected for cognitive machines without any prior assumption about the structure. An example scenario can be imagined for this, and it is not simply the biometric information of a person but different information extracted from some pattern. The system will be able to semantically analyze some handpicked patterns, such as images, multimedia system information, sounds, etc., and then, extract distinctive characteristics or information from them.

9.3.3 Quantum Computing

Configurable extremely short pulses of light can operate computers 100,000 times faster than today's computers by using quantum computing. *Quantum computing* focuses on developing computers based on quantum theory, which deals with the nature and behavior of all the aspects at the quantum level. Recently, companies like Microsoft, IBM and Google have been investing huge amounts of money in quantum computing (García-Torres et al., 2016). In quantum mechanics, data are not allowed to be stored directly. Instead, data are converted into quantum bits before being stored on traditional devices, which requires a lot of information and is very difficult to achieve. So, when these computers become available for commercial use, researchers can easily use DNA computing on those computers for fast processing (Satell, 2016). Recent advances in DNA computing can simplify the aforementioned process to a great extent. By using the concept of quantum computing, the combination of sentimental analysis and DNA storage can be applied in various fields like healthcare.

Many researchers are demanding that DNA computers be even faster than quantum computers. However, their demand is still only on paper. DNA is very stable and it is not affected by high temperatures. Since DNA is a three-dimensional molecule, one extra dimension can be added to store more data per unit area.

9.3.4 DNA Sequence Market

DNA sequencing can be defined as a procedure to determine the order of nucleotides in the DNA molecule. The sequencing market is developing day by day using the technology of cloud computing, where the main emphasis is on the applications and new technologies of DNA sequences, such as pyrosequencing, 454 technology, Massively Parallel Signature Sequencing (MPSS), Sequencing By Synthesis (SBS), Next Generation Sequencing (NGS), etc. (Lin et al., 2011). These technologies enable users to use DNA sequences more efficiently.

A market report by Technavio predicts that the world DNA sequencing product market will grow stupendously at a powerful Compound Annual Growth Rate rate (CAGR). The cost of DNA sequencing will probably contribute to the expansion of this market. The global DNA sequence market was worth $5,000 million in 2016, and it is expected that it will grow to three times this value by 2023.

9.4 Conclusions

Data security plays an important role in IT industries because of the presence of many hackers and malicious users. Every user wants their data to be protected. Currently, DNA computing is one of the most popular approaches to improving data security, due to the complex structure of DNA. DNA computing is also using in many emerging fields, such as Big Data, cloud computing, etc. The DNA-based data storage technique is very efficient because of its high data density, low maintenance cost, durability, etc. In this chapter, many applications of DNA computing have been discussed in detail. Moreover, future trends in DNA computing, such as cognitive computing, quantum computing, etc. have also been presented. Future trends in DNA computing can be highly beneficial for research in this field.

Key Terms and Definitions

Steganography: Steganography is a technique for hiding the existence of data. In DNA steganography, data are hidden by using a DNA sequence. Here, data are first converted into binary form, and then, by using a base pairing rule, the binary data is converted into a DNA sequence. Those pairing rules are always done naturally because the chances for synthesizing the hydrogen bonds between "T" and "A," and between "G" and "C."

Binary coding: In DNA computing, the binary coding rule transforms the DNA bases, namely A, C, G and T into binary codes and vice versa. For example, binary coding rule ((A 00) (C 01) (G 10) (T 11)) or ((A 01) (C 00) (G 11) (T 110)) or some other rule can be used for transforming the DNA bases into binary codes.

Cognitive computing: Cognitive computing can be defined as the simulation of human thinking in a computerized model. Cognitive computing uses pattern recognition, data mining and natural language-processing techniques to mimic the workings of the human brain. Cognitive computing machines continually gather knowledge from the information or data fed into them. The main goal of cognitive computing is to develop an automated digital computer system that can solve difficult problems without human assistance.

Quantum computing: Quantum computing focuses on developing computers based on quantum theory that is able to deal with the nature and behavior of all the aspects at the quantum level. A quantum computer can be defined as a device that executes quantum computing. In quantum mechanics, data are not allowed to be stored directly. In digital computers, data are saved in binary form (0 or 1). On the other hand, in quantum computers, quantum bits are used to store the data. Recently, companies like Microsoft, IBM and Google have been investing huge amounts of money in quantum computing.

References

Abbasy, M. R. and Shanmugam, B. (2011). Enabling data hiding for resource sharing in cloud computing environments based on DNA sequences. In *Proceedings of the 2011 IEEE World Congress on Services*, IEEE, Washington, DC, pp. 385–390.

Adleman, L. M. (1994). Molecular computation of solutions to combinatorial problems. *Science*, 266(5187), 1021–1024.

Arita, M. (2004). Writing information into DNA. In *Aspects of Molecular Computing*. Springer, pp. 23–35.

Chandra, R. (2017). An affective computational model for machine consciousness. *arXiv preprint arXiv:1701.00349*.

Changxiang, S., Zhang, H., Feng, D., Cao Z., and Huang, J. (2007). Survey of information security. *Science in China Series F: Information Sciences*, 50(3), 273–298.

Deka, G. C. and Das, P. K. (2014). An overview on the virtualization technology. *Handbook of Research on Cloud Infrastructures for Big Data Analytics*, IGI Global. DOI: 10.4018/978-1-4666-5864-6.ch012.

García-Torres, M., Gómez-Vela, F., Melián-Batista, B., and Moreno-Vega, J. M. (2016). High-dimensional feature selection via feature grouping: A variable neighborhood search approach. *Information Sciences*, 326, 102–118.

Goldman, N., Bertone, P., Chen, S., Dessimoz, C., LeProust, E. M., Sipos, B., and Birney, E. (2013). Towards practical, high-capacity, low-maintenance information storage in synthesized DNA. *Nature*, 494, 77–80.

Hakami, H. A., Chaczko, Z., and Kale, A. (2015). Review of big data storage based on DNA computing. In *Proceedings of the Asia-Pacific Conference on Computer Aided System Engineering*, IEEE, Quito, Ecuador, pp. 113–117.

Hsu, H. Z. and Lee, R. C. T. (2006). DNA based encryption methods. In *Proceedings of the 23rd Work-Shop on Combinatorial Mathematics and Computation Theory*, Nantou Hsies, Taiwan, pp. 145–150.

Huffman, D. A. (1952). A method for the construction of minimum redundancy codes. *Proceedings of the IRE*, 40(9), 1098–1101.

IDC. (2014). *The Digital Universe of Opportunities: Rich Data and the Increasing Value of the Internet of Things*. Retrieved June 11, 2017, from https://www.emc.com/leadership/digital-universe/2014iview/executive-summary.htm.

Javheri, S. and Kulkarni, R. (2014). Secure data communication and cryptography based on DNA based message encoding. *International Journal of Computer Applications*, 98(16), 35–40.

Johnson, N. F. and Jajodia, S. (1998). Exploring steganography: Seeing the unseen. *IEEE*, 31(2), 26–34.

Kencla, L. and Loebl, M. (2010). DNA-inspired information concealing: A survey. *Computer Science Review*, 4(4), 251–262.

Lin, B., Wang, J., and Cheng, Y. (2011). *Recent Patents and Advances in the Next-Generation Sequencing Technologies*. Retrieved June 11, 2017, from https://www.ncbi.nlm.nih.gov/pmc/articles/PMC3122325/

Monika and Upadhyaya, S. (2015). Secure communication using DNA cryptography with secure socket layer (SSL) protocol in wireless sensor networks. *Procedia Computer Science*, 70, 808–813.

Namasudra, S. (2018). Cloud computing: A new era. *Journal of Fundamental and Applied Sciences*, 10(2), 113–135.

Namasudra, S. (2017). An improved attribute-based encryption technique towards the data security in cloud computing. *Concurrency and Computation: Practice and* Exercise. DOI:10.1002/cpe.4364

Namasudra, S. and Roy, P. (2015). Size based access control model in cloud computing. In *Proceedings of the International Conference on Electrical, Electronics, Signals, Communication and Optimization*, IEEE, Visakhapatnam, India, pp. 1–4.

Namasudra, S. and Roy, P. (2016). Secure and efficient data access control in cloud computing environment: A survey. *Multiagent and Grid Systems – An International Journal*, 12(2) 69–90.

Namasudra, S. and Roy, P. (2017a). Time saving protocol for data accessing in cloud computing. *IET Communications*, 11(10), 1558–1565.

Namasudra, S. and Roy, P. (2017b). A new secure authentication scheme for cloud computing environment. *Concurrency and Computation: Practice and Exercise*, 29(20). DOI: 10.1002/cpe.3864

Namasudra, S. and Roy, P. (2017c). A new table based protocol for data accessing in cloud computing. *Journal of Information Science and Engineering*, 33(3), 585–609.

Namasudra, S. and Roy, P. (2018). PpBAC: Popularity based access control model for cloud computing. *Journal of Organizational and End User Computing*, 30(4), 14–31.

Namasudra, S., Nath, S., and Majumder, A. (2014). Profile based access control model in cloud computing environment. In *Proceedings of the International Conference on Green Computing, Communication and Electrical Engineering*, IEEE, Coimbatore, India, pp. 1–5.

Namasudra, S., Roy, P., and Balamurugan, B. (2017a). Cloud computing: Fundamentals and research issues. In *Proceedings of the 2nd International Conference on Recent Trends and Challenges in Computational Models*, IEEE, Tindivanam, India.

Namasudra, S., Roy, P., Balamurugan, B., and Vijayakumar, P. (2017b). Data accessing based on the popularity value for cloud computing. In *Proceedings of the International Conference on Innovations in Information, Embedded and Communications Systems (ICIIECS)*, IEEE, Coimbatore, India.

Namasudra, S., Roy, P., Vijayakumar, P., Audithan, S., and Balamurugan, B. (2017c). Time efficient secure DNA based access control model for cloud computing environment. *Future Generation Computer Systems*, 73, 90–105.

Sarkar, S., Saha, K., Namasudra, S., and Roy, P. (2015). An efficient and time saving web service based android application. *SSRG International Journal of Computer Science and Engineering (SSRG-IJCSE)*, 2(8), 18–21.

Satell, G. (2016). *Here's How Quantum Computing Will Change The World*. Retrieved June 11, 2017, from https://www.forbes.com/sites/gregsatell/2016/10/02/heres-how-quantum-computing-will-change-the-world/#675fa7ad6d0a.

Shah, S., Limbachiya, D., and Gupta, M. K. (2013). DNACloud: A potential tool for storing big data on DNA. *Emerging Technologies*. DOI: arXiv:1310.6992

Shiu, H. J., Ng, K. L., Fang, J. F., Lee, R. C. T., and Huang, C. H. (2010). Data hiding methods based upon DNA sequences. *Information Sciences*, 180(1), 2196–2208.

Somnath, S. M., Bhattacharya, S., Islam, M. A., and Islam, M. L. (2010). DNA computation: Applications and perspectives, *Journal of Proteomics and Bioinformatics*, 3(7), 234–243.

Sureshraj, D. and Bhaskaran, V. M. (2012). Automatic DNA sequence generation for secured cost-effective multi-cloud storage. In *Proceedings of the International Conference on Software Engineering and Mobile Application Modelling and Development*, pp. 1–6.

Tanaka, K., Okamoto, A., and Saito I., (2005). Public-key system using DNA as a one-way function for key distribution, *Biosystems*, 81(1), 25–29.

Terec, R., Vaida, M. F., Alboaie, L., and Chiorean, L. (2011). DNA security using symmetric and asymmetric cryptography. *International Journal on New Computer Architectures and their Applications (IJNCAA)*, 1(1), 34–51.

Wang, B., Xie, Y., Zhou, S., Zhou, C., and Zheng, X. (2017). Reversible data hiding based on DNA computing. *Computational Intelligence and Neuroscience*, 2017, 1–9.

Wikipedia. (2018). DNA. Retrieved February 17, 2018, from https://simple.wikipedia.org/wiki/DNA.

Zhang, S. and Gao, T. (2015). A novel data hiding scheme based on DNA coding and module-N operation. *International Journal of Multimedia and Ubiquitous Engineering*, 10(4), 337–344.

References for Advanced/Further Reading

Chang, W. L. (2012). Fast parallel DNA-based algorithms for molecular computations: Quadratic congruence and factoring integers. *IEEE Transactions on NanonBioscience*, 11(1), 62–69.

Danziger, M. and Henriques, M. A. A. (2012). Computational intelligence applied on cryptography: A brief review. *IEEE Latin America Transactions*, 10(3), 1798–1810.

Singh, S. (2018, March 29). Andhra Pradesh adopts blockchain technology for DNA sequencing of citizens. Retrieved March 30, 2018, from https://techobserver.in/article/egov/andhra-pradesh-signs-mou-with-genomics-firm-shivom-to-use-blockchain-for-predictive-medicine-privacy-of-citizens-dna-data.

Index

E

F

G

H

I

K

L

M

N

U

V

Z

Advances of DNA Computing in Cryptography